A – Z
Maths
Games

Karen M. Breitbart

Brilliant Publications

A special thanks to ...

The Citibank Success Fund for providing the grant that helped to purchase the items necessary for *A – Z Maths Games.*

Mr Robert F. Morgan, my Principal, for allowing me the freedom to try new things in my classroom.

Most importantly, my loving husband, Gregg, for always encouraging me to succeed.

About the author

Karen M. Breitbart received a Bachelor of Science degree in Education at Louisiana State University and a Master's degree in Early Childhood Education at Florida International University. She has been teaching for ten years.

Publisher's information

Published by:
Brilliant Publications
The Old School Yard
Leighton Road   Tel: **01525 222844**
Northall   Fax: **01525 221250**
Dunstable   Website: **www.brilliantpublications.co.uk**
Bedfordshire LU6 2HA E-mail: **sales@brilliantpublications.co.uk**

A to Z Maths Games by Karen M. Breitbart
Illustrated by Marilynn G. Barr
© Monday Morning Books Inc.
Originally published in the USA in1997 by Monday Morning Books Inc.
ISBN 1 897675 79 8
First published in the UK 2001
10 9 8 7 6 5 4 3 2 1
Printed in Malta by Interprint Ltd

The right of Karen M. Breitbart to be identified as the author of this work has been asserted by her in accordance with the Copyright, Design and Patents Act 1988.

Contents

Introduction

A – Z Maths Games reinforces basic maths skills and concepts. The games in this resource will help children build a strong maths foundation on which to base future learning experiences.

The A – Z Maths Games programme assists children in achieving success by providing them with tools for hands-on exploration and manipulation. These high-interest, self-motivating games are intended for individual use.

All games in this resource have been 'kid-tested' and approved! The children loved to learn this way and looked forward to learning maths skills. Each child was able to work at his or her own pace, and they progressed, regardless of their varying abilities.

After you have made the games, you will have them for years. Since only one child plays with each game at a time, you avoid the problem of losing parts.

I hope that you and your students enjoy these games as much as my students and I enjoyed creating them!

In the classroom

A good technique for incorporating A – Z Maths Games in the classroom is to set up a Maths Centre containing three to five of the games per week. There are many ways to organise the use of these games.

Book links

The books listed are available from good bookshops and through Amazon via the internet at www.amazon.co.uk.

Brilliant Publications – www.brilliantpublications.co.uk
A – Z Maths Games by Karen M. Breitbart

Making the games

♦ Each game contains directions for making and using motivating hands-on games in a classroom setting.

♦ For each game, all patterns are included. Some games may require additional readily available materials.

♦ Colour the patterns as desired.

♦ To keep your games in the best possible shape, laminate the pieces. If you do not have access to a laminating machine, you can also use clear sticky-back plastic to cover the pieces.

♦ Consider labelling patterns on the back for children to self-check their work.

Setting up a Maths Centre area

♦ Find five different coloured plastic baskets (or use five of the same coloured baskets, labelled with different coloured sugar paper).

♦ Label the table you've chosen for your Maths Centre area with five sugar paper circles. These circles should be the same colour as the baskets or the basket labels.

♦ Permanently attach these circles to the table using tape and then cover them with clear contact paper. The baskets will always be stored on top of the corresponding coloured circle.

♦ Select five games that you would like your students to work with. Put one game in each basket.

Preparing the Maths Centre groups

♦ Divide your children into groups of five.

♦ Assign each child a colour, so that there is one 'red', 'yellow', 'blue', 'orange' and 'green' child in each group. (Your colours may vary, depending on what colour baskets you use.)

♦ Set up your groups before the children come in to school and write the children's names on the coloured circles that are permanently attached to the Maths Centre table. (Write the names with a permanent marker.)

♦ Allow the children to sit with their groups and decide on a group name. The children will stay together all year.

♦ Tell the children what their colours are.

How to use the games

♦ On Monday, introduce these games to the children. Directions are included for you to read to the children.

♦ First, model each game step by step, from setting up to playing to tidying up.

♦ After you have modelled each game, choose one or two children to demonstrate how to set up, play with, and tidy up the game.

♦ At Centre Time, have the groups rotate through all of the centres in the room, including the Maths Centre. When a group gets to the Maths Centre, each child will take and use the game that is in his or her colour basket.

♦ At Tidy Up Time, children will replace their game in the basket and place it on the corresponding colour, making sure the games are ready for the next group.

♦ At the end of the day, rotate the games from left to right. The game in the red basket will move to the blue basket, the blue to the orange, and so on, with the last game in the row becoming the first. By rotating the games, the children will find a different game in their basket each day and you will be able to keep track of who is playing which game. By the end of the week, all of your children will have had time to play with each game.

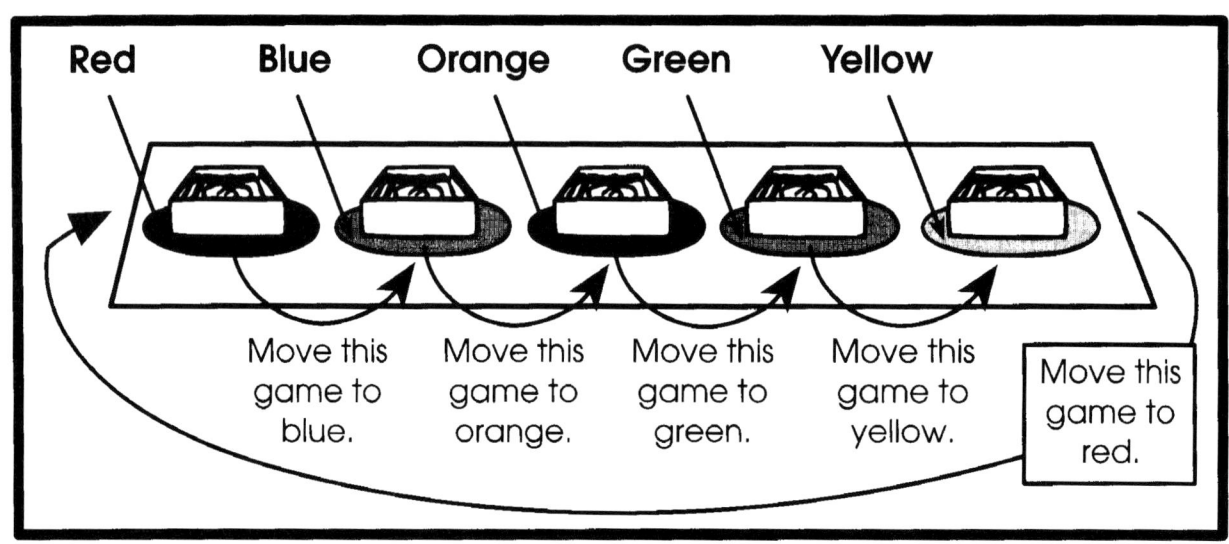

Brilliant Publications – www.brilliantpublications.co.uk
A – Z Maths Games by Karen M. Breitbart

Alligators

Objective:
Children will make number sets from 1 to 5.

Materials:
Alligators (page 8); Fish (page 9); scissors; 5 split pins; hole punch; crayons or markers; resealable bag.

How to make the game:
♦ Duplicate the pages of alligators and fish, colour, laminate and cut out.
♦ Punch a hole in each of the alligators' top and bottom jaws (as indicated).
♦ Attach the alligators jaws together using the split pins.
♦ Store all game pieces in the resealable bag.

How to play the game:
♦ Take the pages of alligators and fish out of the bag.
♦ Each alligator has a number on its hat. Place the alligators in a line from 1 to 5.
♦ Match sets of fish to the number on each alligator. For example, for the first alligator, you will have a set of one Fish.
♦ 'Feed' the fish to the alligators.
♦ When you have finished, ask your teacher to check your work. Then place all game pieces back in the bag.

Book link:
Alligators All Around: An Alphabet by Maurice Sendak (HarperCollins, 1991). Alligators perform activities for each letter of the alphabet.

Alligators

Brilliant Publications – www.brilliantpublications.co.uk
A – Z Maths Games by Karen M. Breitbart

Fish

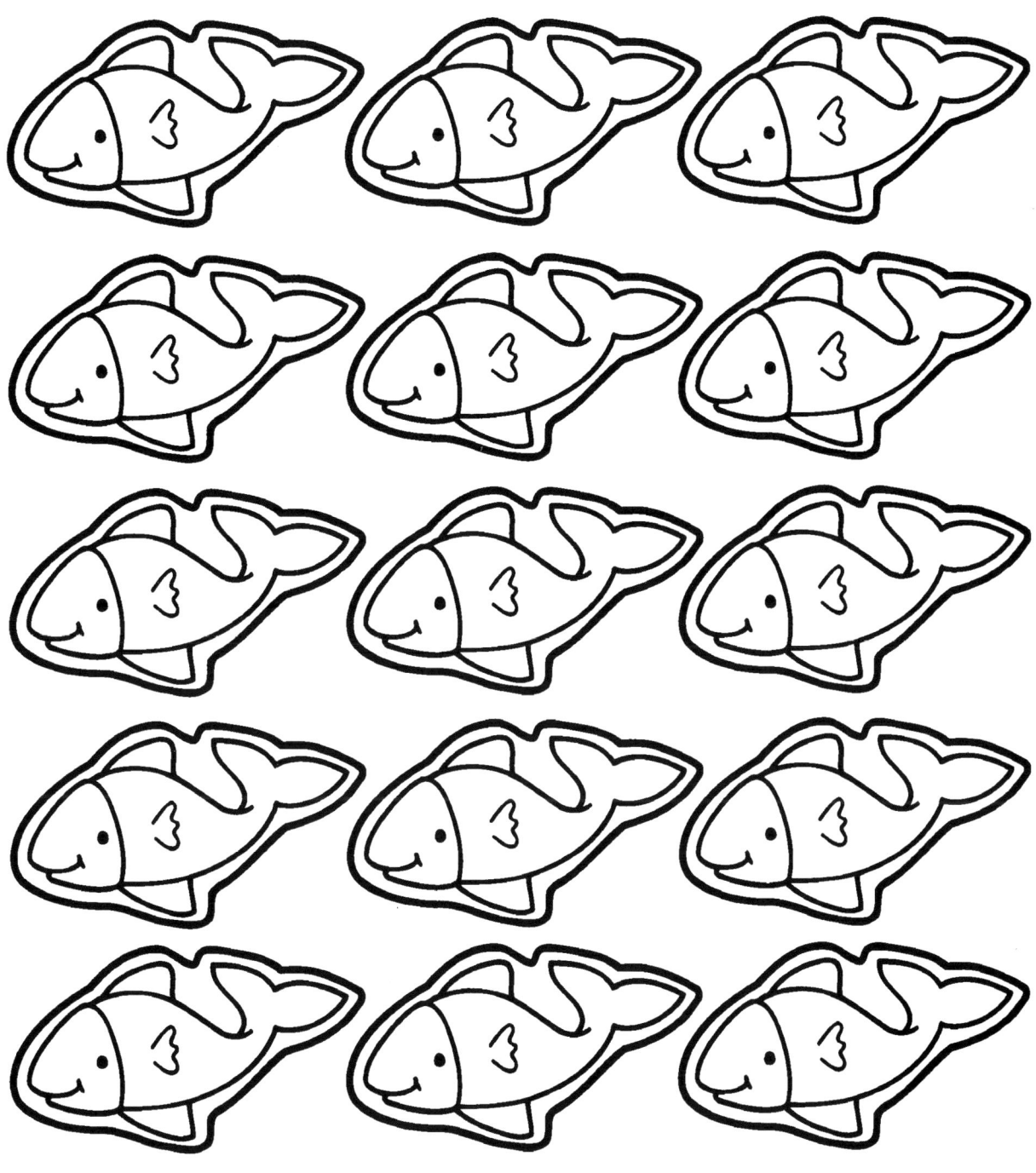

Aeroplane maths

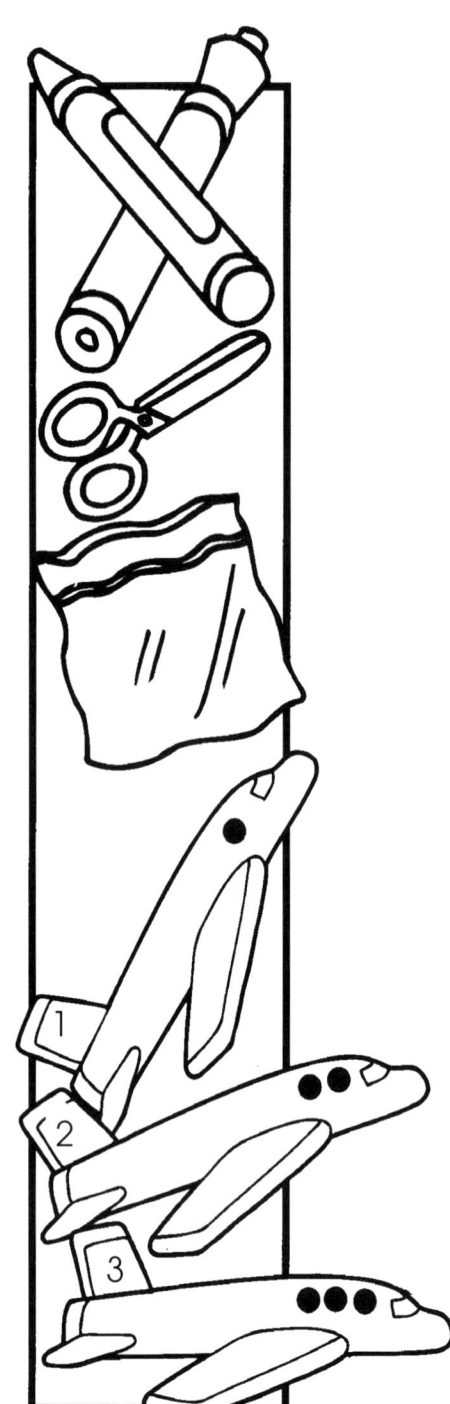

Objective:

Children will match dots to the correct numerals.

Materials:

Aeroplane bodies (page 11); Aeroplane tails (page 12); crayons or markers; scissors; resealable bag.

How to make the game:

♦ Duplicate the aeroplane bodies and tails, colour, laminate and cut out.
♦ Store all game pieces in the resealable bag.

How to play the game:

♦ Take the game pieces out of the bag. Line the aeroplane bodies in one row and the aeroplane tails in another.
♦ The aeroplane bodies have dots, and the tails have numbers.
♦ Place each aeroplane tail on the correct aeroplane body.
♦ When you have finished, ask your teacher to check your work. Then place all game pieces back in the bag.

Book link:

This Plane by Paul Colliatt (Farrar Straus Giroux, 2000).

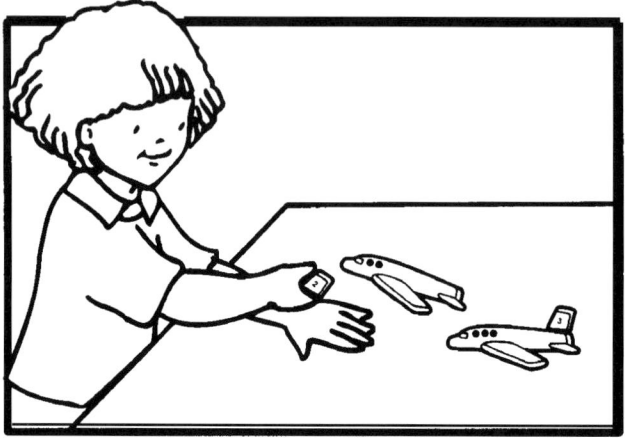

Brilliant Publications – www.brilliantpublications.co.uk
A – Z Maths Games by Karen M. Breitbart

Aeroplane bodies

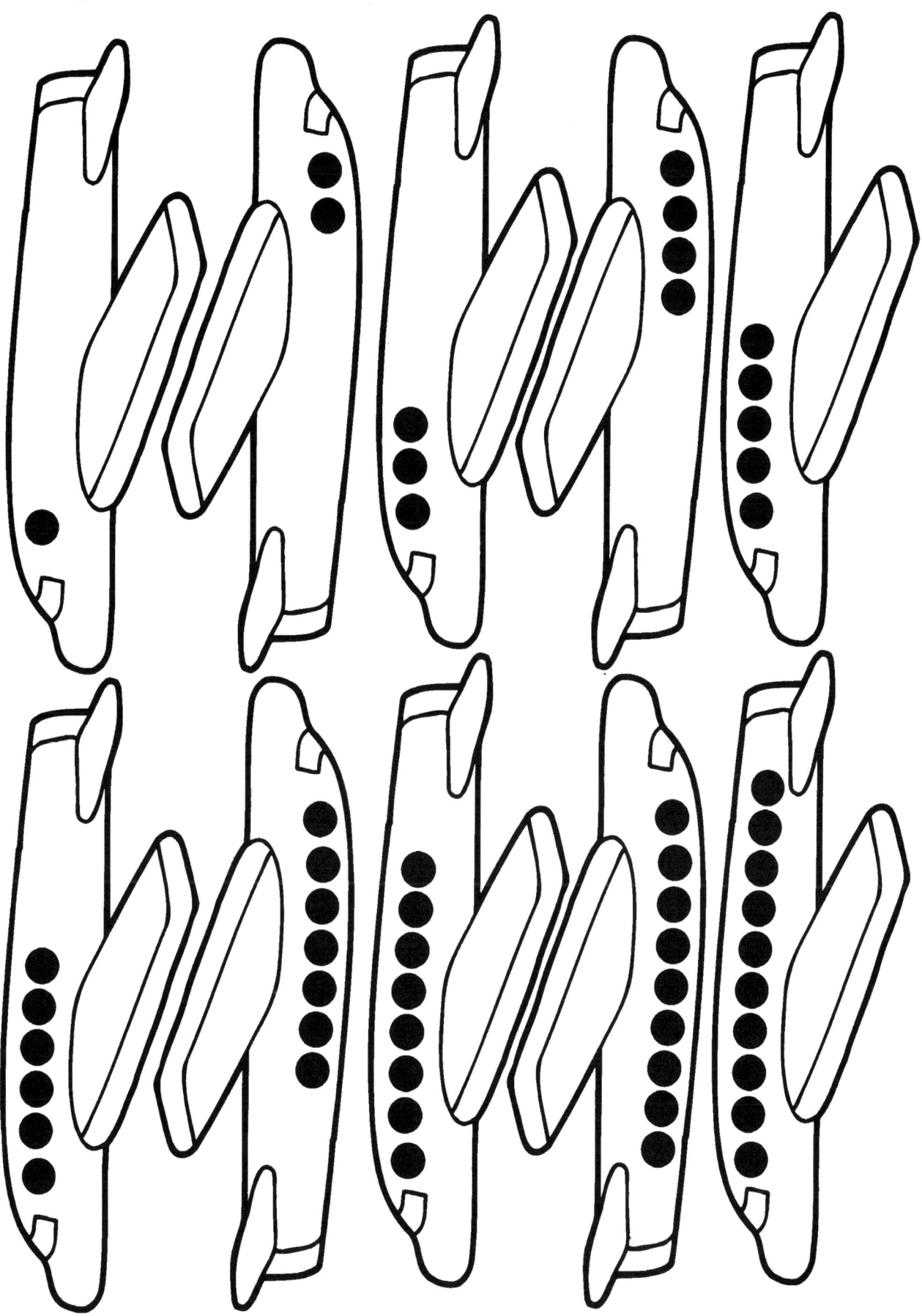

Aeroplane tails

Brilliant Publications – www.brilliantpublications.co.uk
A – Z Maths Games by Karen M. Breitbart

Bean toss

Objective:
Children will count and record numbers on a graph.

Materials:
Happy bean graph (page 14); crayons; permanent marker; 5 large butter beans; large envelope; resealabe bag.

How to make the game:
♦ Duplicate the happy bean graph. Make one copy per child.
♦ Use the permanent marker to draw a happy face on one side of each butter bean.
♦ Store the beans and a crayon in the resealable bag, and store the happy bean graphs in a large envelope.

How to play the game:
♦ Take one happy bean graph out of the envelope.
♦ Take the beans out of the bag and toss them on to the table.
♦ Sort the beans into two piles: one with the happy faces up and one with the blank sides up.
♦ Count the number of happy faces and record this on the graph by writing the numeral in the correct column.
   Toss the beans four times.
♦ When you have finished, give your graph to your teacher.
♦ Then place the beans and the crayon in the resealable bag.

How many?	HAPPY BEAN GRAPH NAME					
					3	
0	1	2	3	4	5	

Book link:
Jack and the Beanstalk by Steven Kellogg (Morrow Junior Books, 1991). Kellogg retells the story in black and white with green additions.

Happy bean graph

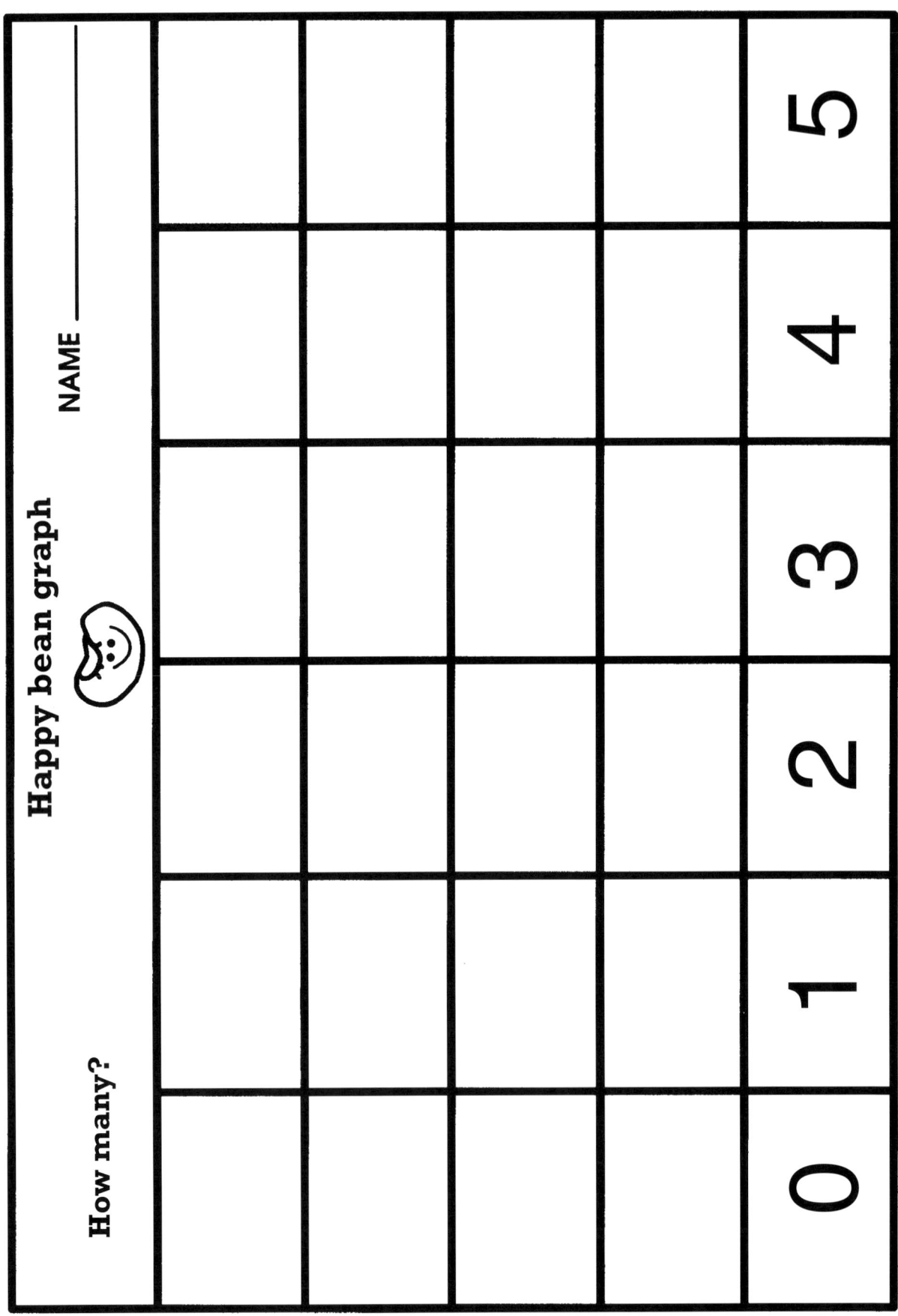

Brilliant Publications – www.brilliantpublications.co.uk
A – Z Maths Games by Karen M. Breitbart

Block numbers

Objective:
Children will use blocks to make number sets from 1 to 10.

Materials:
Number cards (page 16); 55 small wooden blocks or beads; scissors; 2 large resealable bags.

How to make the game:
♦   Duplicate the number cards, laminate and cut apart.
♦   Store the number cards in one bag and the blocks in another.

How to play the game:
♦   Line the number cards in order from 1 to 10
♦   Place the correct number of blocks in front of each card. For example, place one block by the card that has the number 1.
♦   When you have finished, ask your teacher to check your work. Then place the blocks back in one bag and the number cards in the other.

Option:
Children can use blocks to outline the numbers on enlarged number cards.

Book link:
But Not the Hippopotamus by Sandra Boynton (Little Simon, 1982). Simple story with nice rhyming.

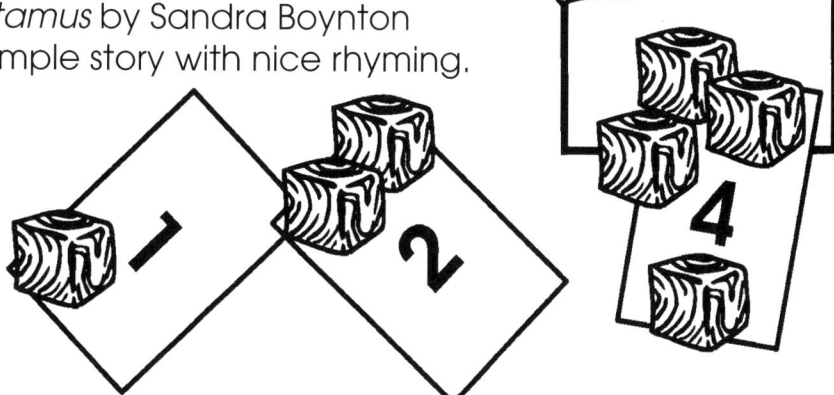

Number cards

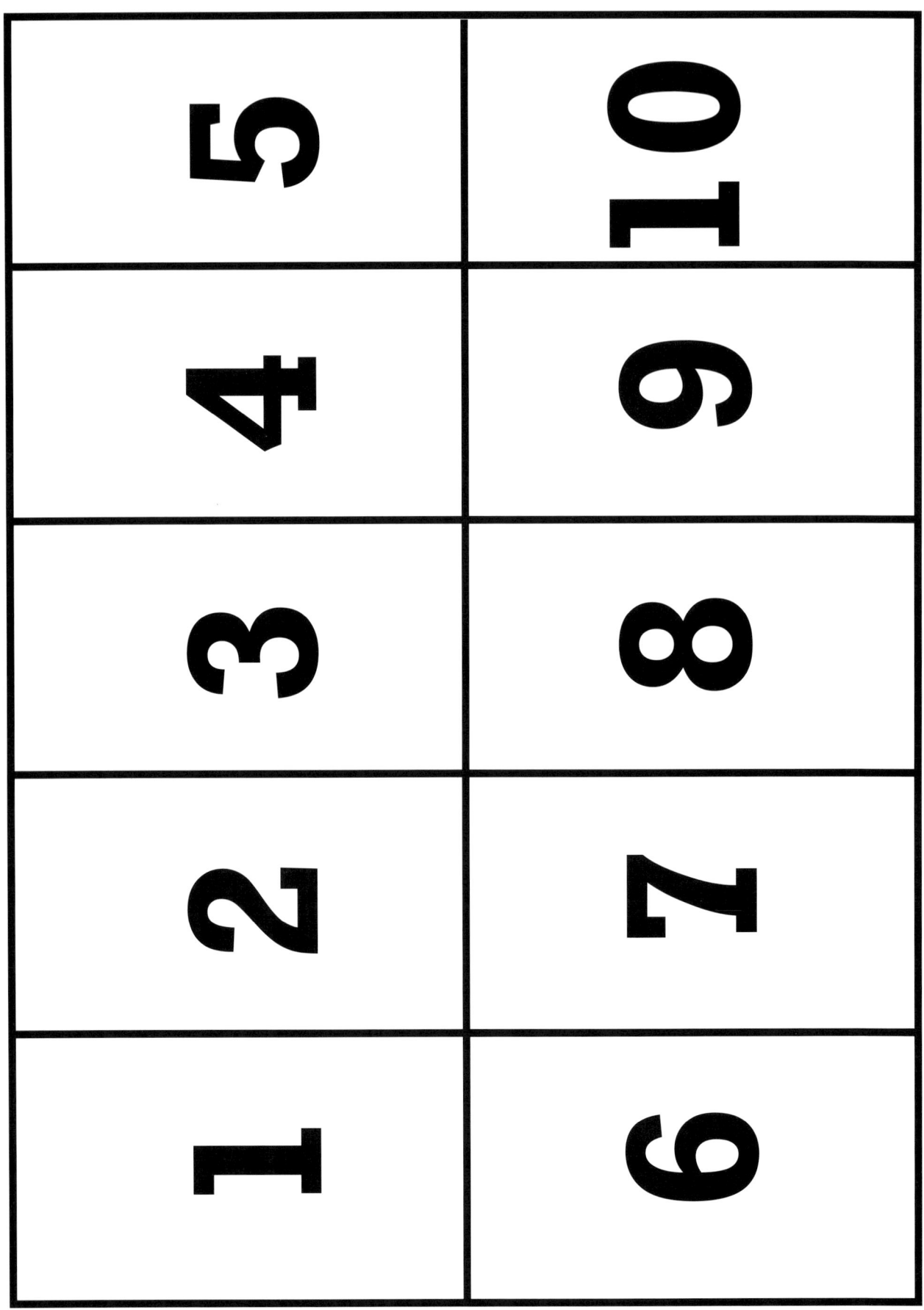

Brilliant Publications – www.brilliantpublications.co.uk
A – Z Maths Games by Karen M. Breitbart

Calico cats

Objective:
Children will practise writing the numbers 1 to 9.

Materials:
Cat (page 18); 9 different colours of crayons or markers; resealable bag; large envelope.

How to make the game:
♦ Duplicate one copy of the cat per child.
♦ Store the cats in a large envelope and the crayons or markers in the resealable bag.

How to play the game:
♦ Take one copy of the cat out of the envelope.
♦ Find the small numbers in each section of the cat.
♦ Use a different colour crayon or marker to write each number over and over again in each section.
♦ When you have finished, place the crayons or markers back in the bag, and give your picture to the teacher.

Note:
Post the finished cat pictures on a 'Calico cat' bulletin board.

Option:
Use masking tape labels to number the crayons from one to nine. Children can colour their cats using the numbered crayons.

Book link:
The Antique Shop Cat by Leslie Baker (Little, Brown, 2000). A calico cat makes an 'unauthorized' visit to an antique shop.

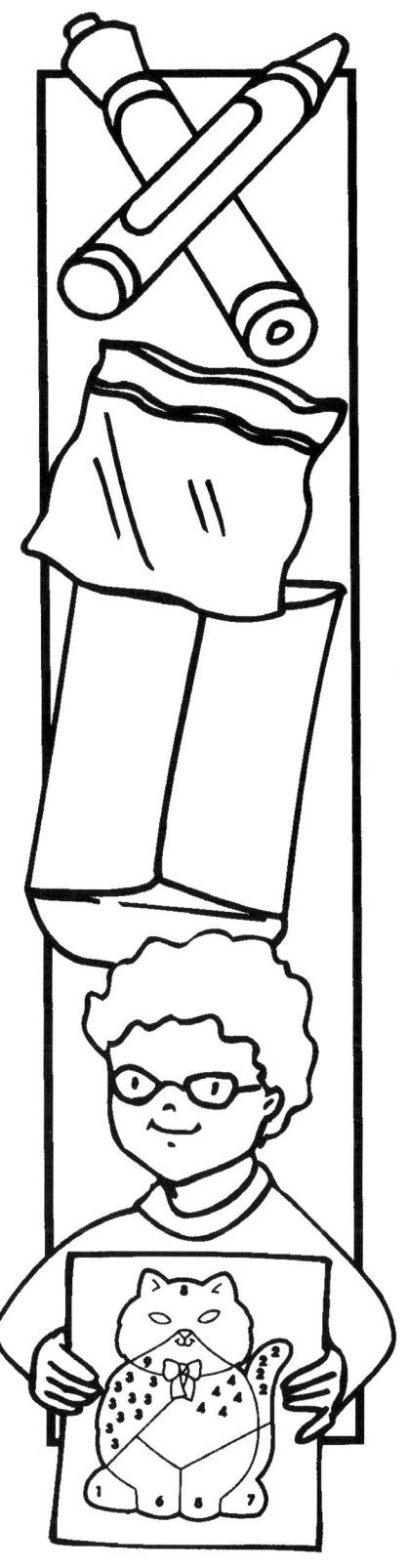

Cat

Brilliant Publications – www.brilliantpublications.co.uk
A – Z Maths Games by Karen M. Breitbart

Clocks

Objective:
Children will practise telling the time.

Materials:
Clock and clock hands (page 20); Clock cards (page 21); heavy paper; split pin; crayons or markers; hole punch; scissors; large envelope; resealable bag.

How to make the game:
- Duplicate the clock and clock hands onto heavy paper, colour, laminate and cut out.
- Punch a hole in the centre of the clock and attach the clock hands using the split pin.
- Duplicate the clock cards, colour, laminate, and cut out.
- Store the clock in a large envelope and the clock cards in the resealable bag.

How to play the game:
- Take the clock out of the envelope and point the hands to 12.
- Take the clock cards out of the bag and place them in a row.
- One by one, look at the clock on each clock card. Then move the hands on the clocks that they are the same as the clock on each card. Read the times aloud.
- When you have finished, place the clock back in the envelope and place the clock cards in the bag.

Book link:
The Bad-tempered Ladybird by Eric Carle (Puffin, 1982). A grouchy ladybird goes through a whole day looking

for a fight.

Clock and clock hands

Brilliant Publications – www.brilliantpublications.co.uk
A – Z Maths Games by Karen M. Breitbart

Clock cards

Domino numbers

Objective:
Children will use the dominoes to form the numbers 1 to 10.

Materials:
10 pieces of thick card; crayons or markers; dominoes; large envelope; resealable bag.

How to make the game:
♦ Write one number (from 1 to 10) on each piece of thick card.
♦ Store the number cards in the large envelope and the dominoes in the resealable bag.

How to play the game:
♦ Spread the number cards face up on the floor or on a table.
♦ Use the dominoes to trace the numbers. Match the dots on the first domino with another domino piece that has the same number of dots. Continue matching as you trace the number.
♦ When you have traced the numbers on all of the cards, ask your teacher to check your work. Then place the cards in the envelope and the dominoes in the bag.

Brilliant Publications – www.brilliantpublications.co.uk
A – Z Maths Games by Karen M. Breitbart

In the dog house

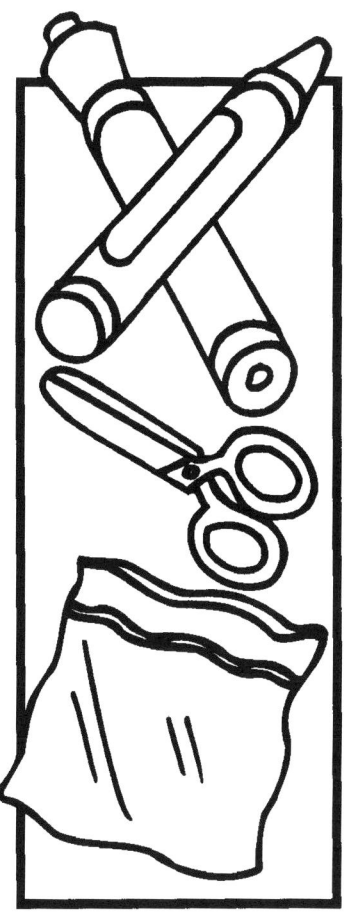

Objective:
Children will match dots to the correct numerals.

Materials:
Dalmatians (page 24); Kennels (page 25); crayons or markers; scissors; resealable bag.

How to make the game:
♦ Duplicate the dalmatians, laminate and cut out.
♦ Duplicate the kennels, colour, laminate and cut out.
♦ Store all game pieces in the resealable bag.

How to play the game:
♦ Take the kennels and dalmatians out of the bag.
♦ Each kennel has a number on it. Line the kennels in a row from 1 to 10.
♦ Count the spots on each dalmatian.
♦ Match each dalmatian to the correctly numbered kennel.
♦ When you have finished, ask your teacher to check your work. Then place all game pieces back in the bag.

Book link:
Spot Counts from 1 to 10 by Eric Hill (Putman Publishing Group, 1989).

Dalmatians

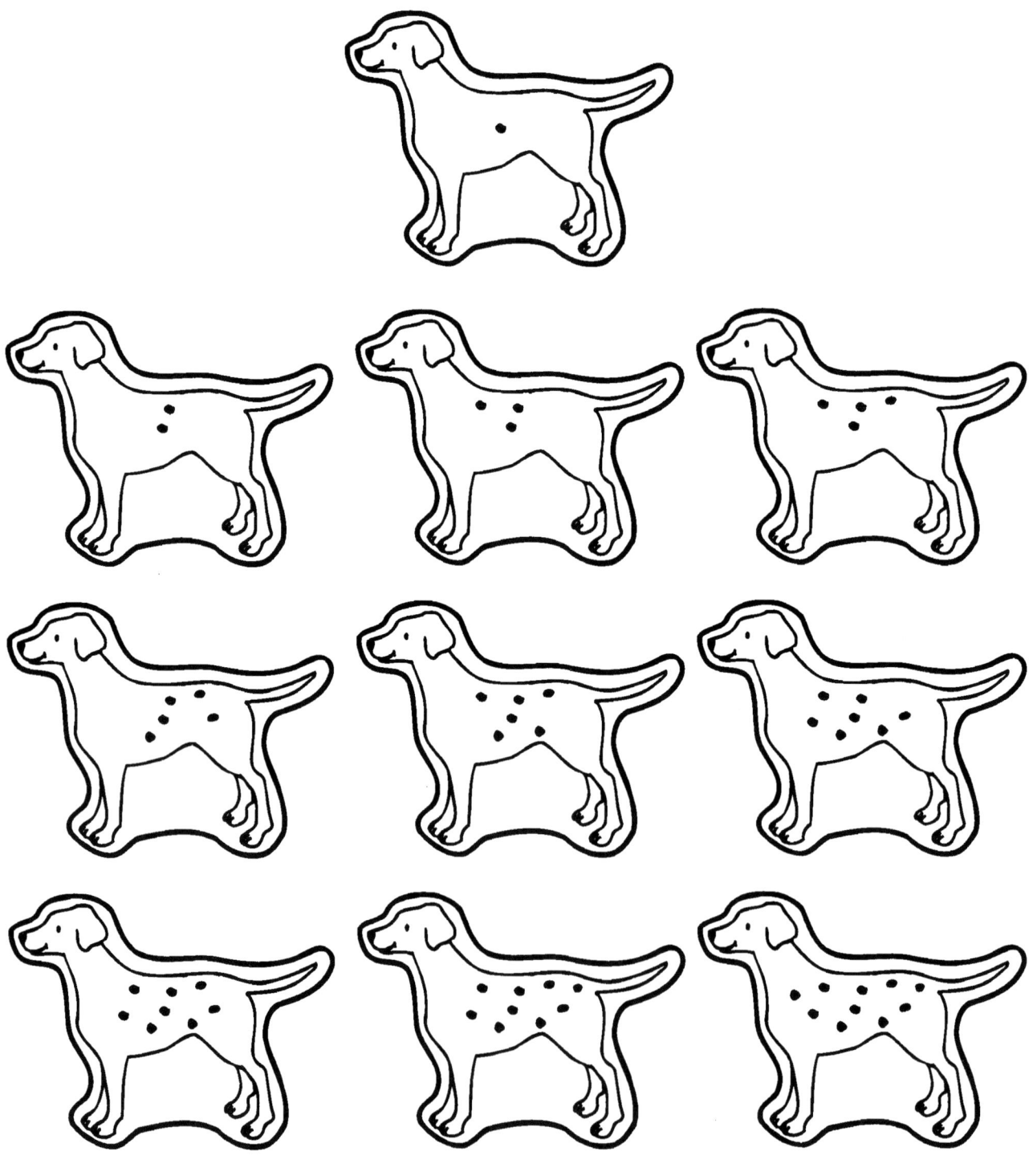

Brilliant Publications – www.brilliantpublications.co.uk
A – Z Maths Games by Karen M. Breitbart

Kennels

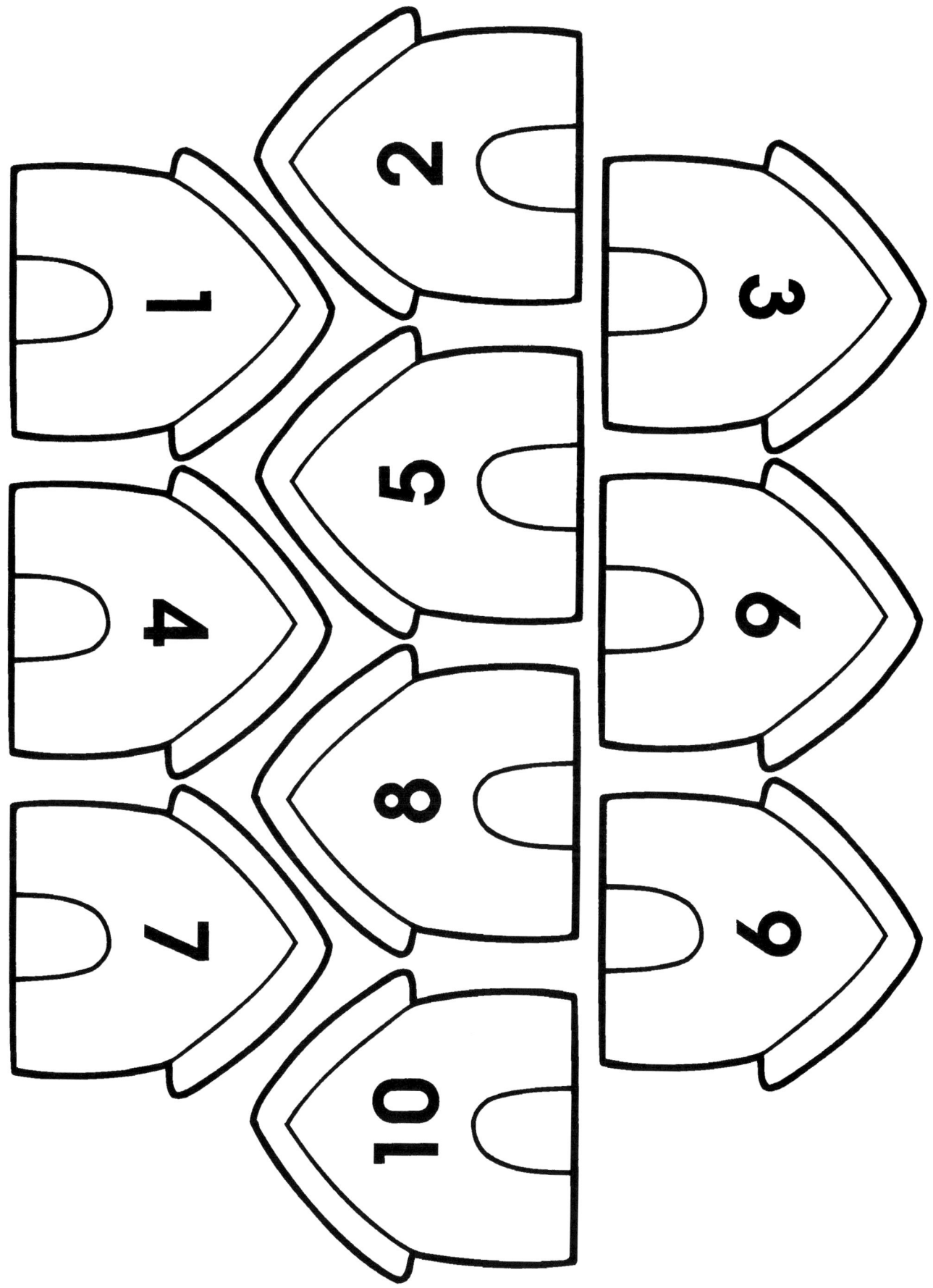

Eggs are hatching

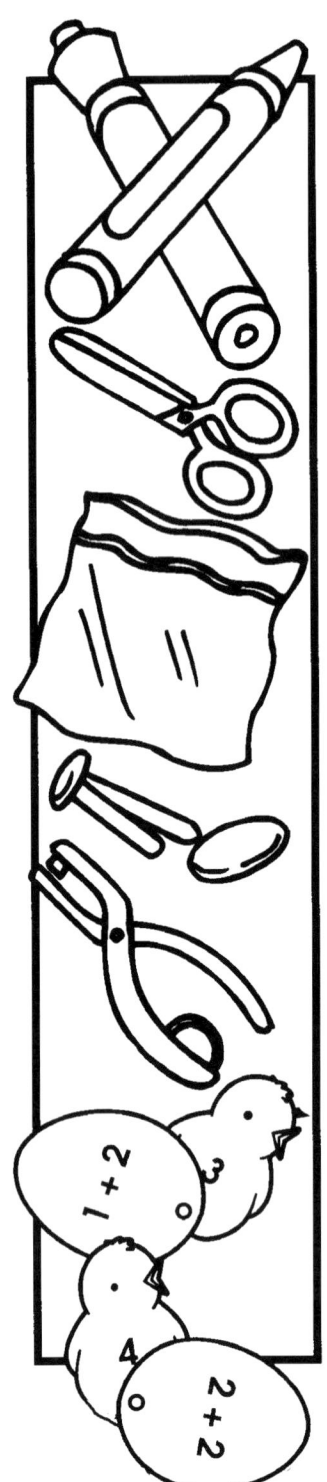

Objective:
Children will use counters to solve addition problems.

Materials:
Eggs (page 27); Chicks (page 28); crayons or markers; scissors; hole punch; 6 split pins; counters; resealable bag.

How to make the game:
♦ Duplicate the eggs, laminate and cut out.
♦ Duplicate the chicks, colour, laminate and cut out.
♦ Punch a hole in each egg and each chick (as indicated).
♦ Use a split pin to attach each chick to the back of the correct egg.
♦ Store all game pieces and counters in a resealable bag.

How to play the game:
♦ Spread the eggs in a row in front of you.
♦ Use the counters to solve the problem on each egg.
♦ Check your work by sliding the chick up from behind each egg. The number on the chick should match your answer.
♦ When you have finished, place all game pieces in the bag.

Option:
Don't attach the chicks to the eggs. Ask the children to match the correct chick to each egg.

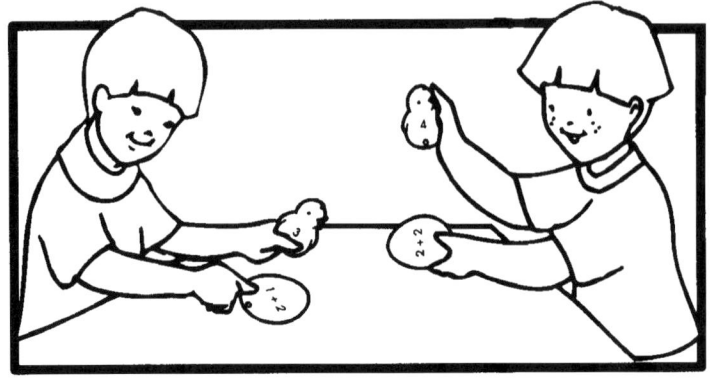

Book link:
The Chick That Wouldn't Hatch Claire Daniel, illustrated by Lisa Campbell Ernst (Green Light Readers, Harcourt Trade Publishing, 1999).

Eggs

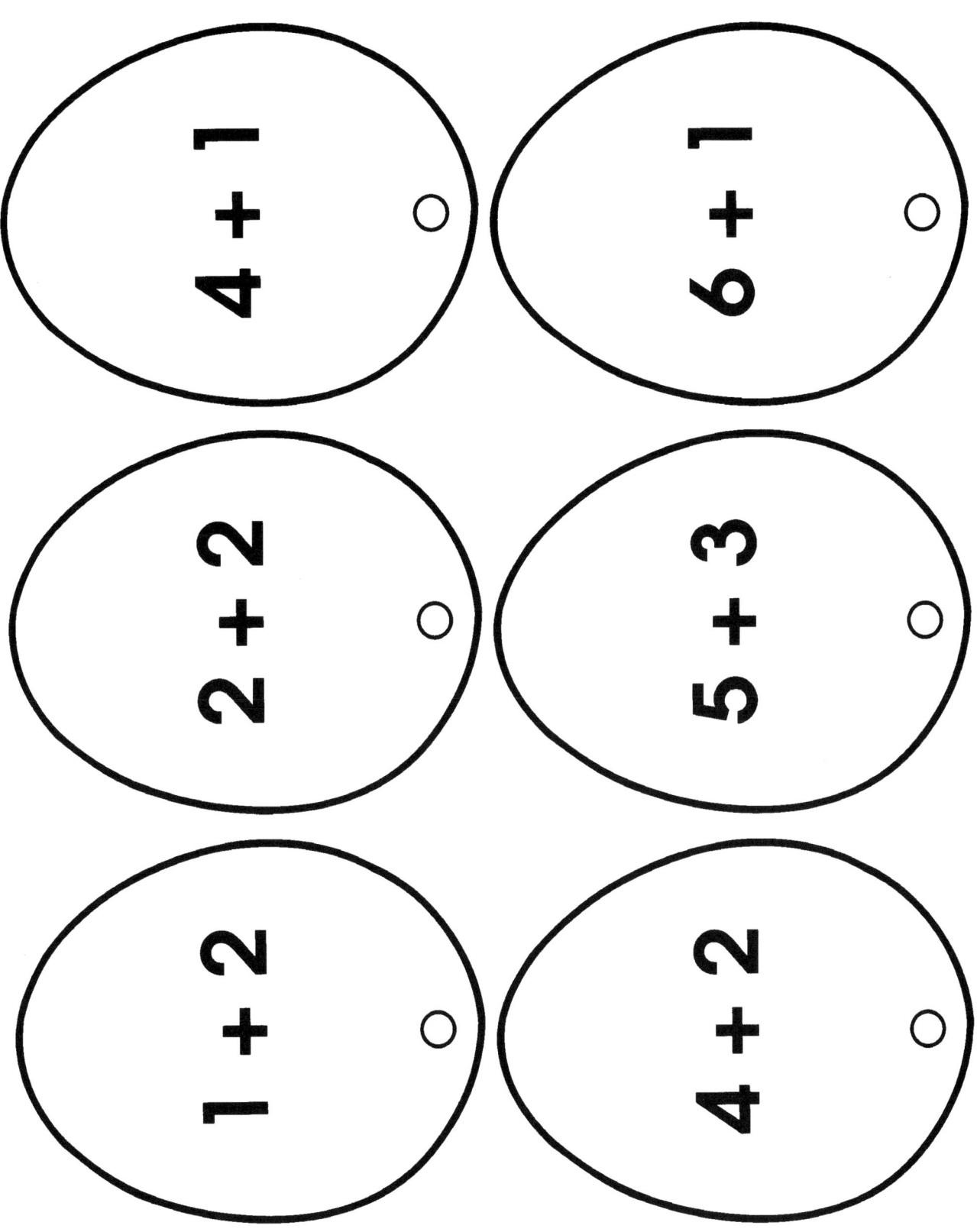

Chicks

Brilliant Publications – www.brilliantpublications.co.uk
A – Z Maths Games by Karen M. Breitbart

Egg box game

Objective:
Children will practise writing the numbers 1 to 6.

Materials:
Record sheet (page 30); six-sectioned egg box; permanent marker; crayon; marble; large envelope; resealable bag.

How to make the game:
♦ Duplicate one copy of the record sheet per child.
♦ Write a number from 1 to 6 in each section of the egg box.
♦ Store the crayon and marble in the bag, inside the box.
♦ Store the record sheets in the large envelope.

How to play the game:
♦ Take one record sheet out of the envelope, and take the marble and crayon out of the bag.
♦ Place the marble in the egg box and close the top.
♦ Shake the egg box with the marble inside.
♦ Open the egg box to see where the marble landed.
♦ Record the number of the egg box section in the correct row by writing the number on one of the egg patterns.
♦ Continue shaking the marble and marking the number on the record sheet until you fill up one row.
♦ When you have finished, give your record sheet to your teacher. Then place the marble and the crayon in the bag and the bag inside the egg box.

Option:
Ask the children to compare their finished sheets. Make a class bar graph showing which number 'won'.

Book link:
A Caribbean Counting Book by Faustin Charles, illustrated by Roberta Anderson (Barefoot Books, 1997). This is a delightful collection of rhymes to be chanted.

Record sheet

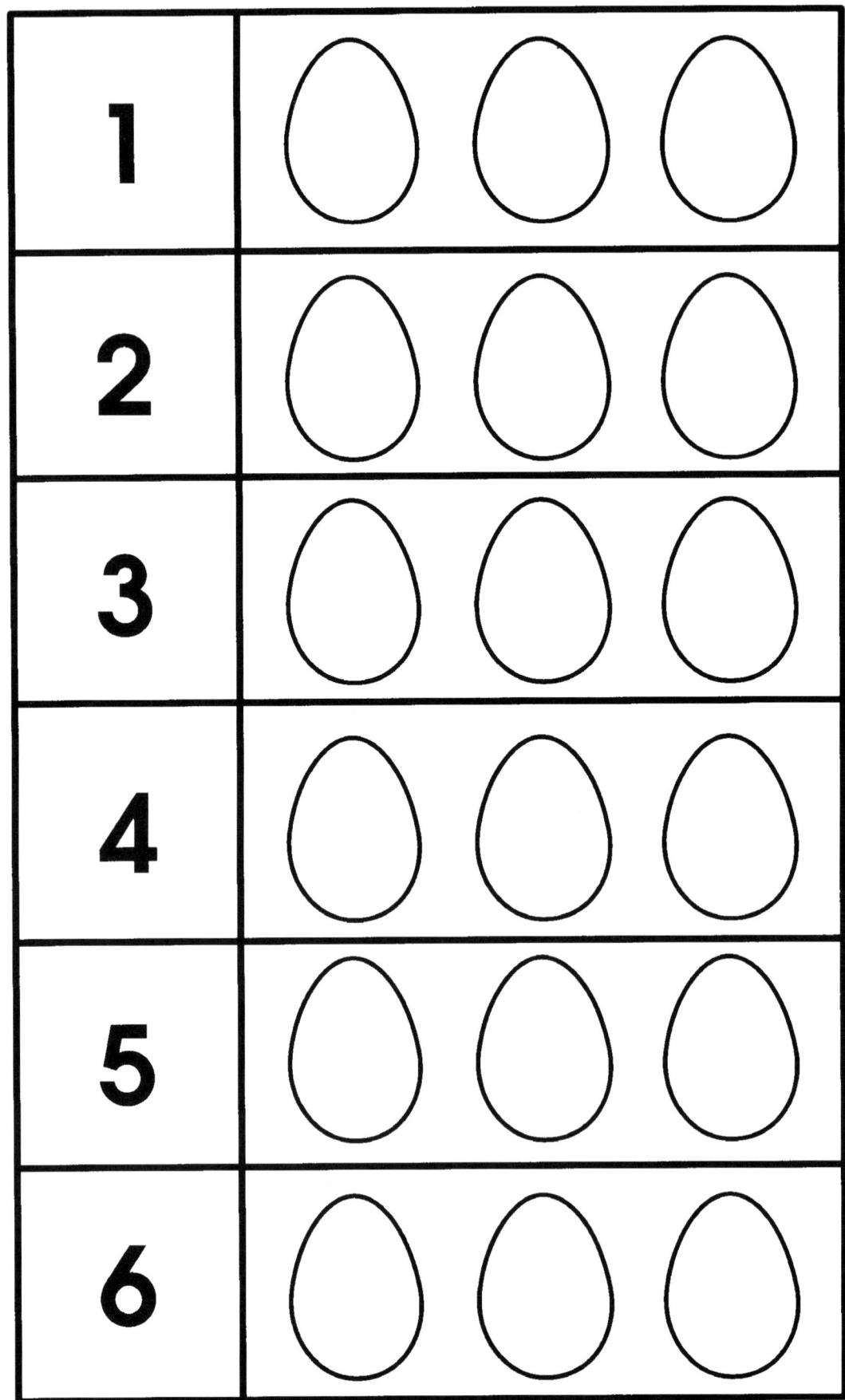

Brilliant Publications – www.brilliantpublications.co.uk
A – Z Maths Games by Karen M. Breitbart

Fish game

Objective:
Children will make sets of the numbers 1 to 10.

Materials:
Number fish (page 32); crayons or markers; small magnet with a hole in the centre; ruler or stick; wool; 10 paper clips; scissors; counters; resealable bag.

How to make the game:
♦ Duplicate the page of number fish, colour, laminate and cut out.
♦ Attach a paper clip to each number fish.
♦ Attach a length of wool to the end of the ruler or stick.
♦ Tie the magnet to the end of the piece of wool.
♦ Store the number fish and counters in the resealable bag.

How to play the game:
♦ Spread the number fish on the floor.
♦ Go fishing with the fishing rod!
♦ After catching a fish, look at the number on the front of it.
♦ Count out the same amount of counters and place them next to the fish.
♦ After catching all of the fish, place the number fish and the counters back in the bag.

Book link:
One Fish, Two Fish, Red Fish, Blue Fish by Dr Seuss (HarperCollins, 1984). This rhyming classic is perfectly suited for this counting game.

Number fish

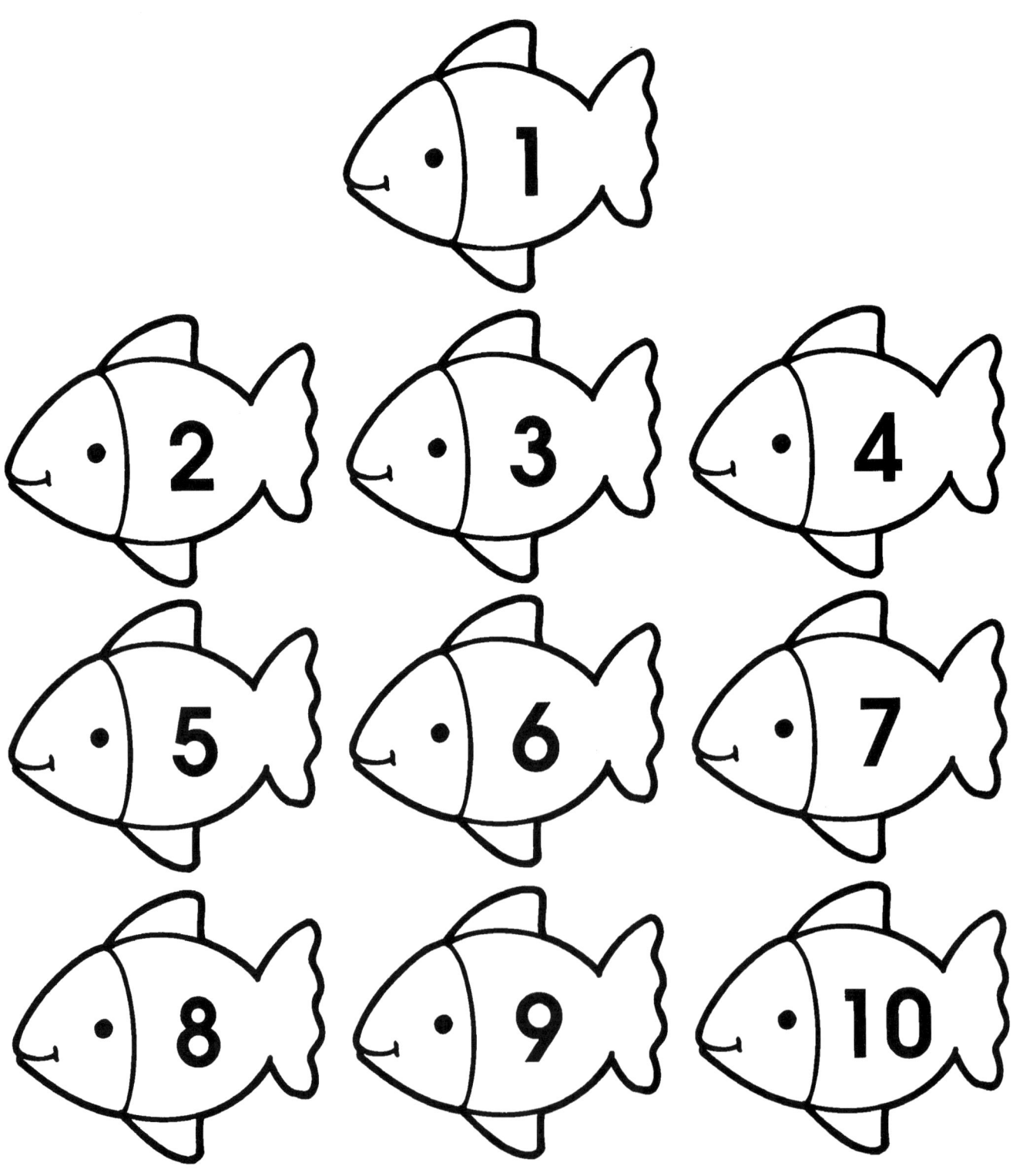

Brilliant Publications – www.brilliantpublications.co.uk
A – Z Maths Games by Karen M. Breitbart

Frog graph

Objective:
Children will practise sorting, counting and graphing.

Materials:
Frogs (page 34); Frog graph (page 35); crayons or markers; scissors; large envelope; resealable bag.

How to make the game:
♦ Duplicate the frogs, colour, laminate and cut out. Make as many copies of the frogs as you'd like, up to ten of each.
♦ Duplicate one copy of the frog graph per child.

How to play the game:
♦ Use the crayon to mark the total number of each type of frog on the frog graph.
♦ When you have finished, place the crayon and the frogs in the bag and give your graph to your teacher.

Option:
Bind the frog graphs in a classroom graph book.

Book link:
I Love to Eat Bugs by John Strejan (Koneman UK, 1998). A pop-up book.

Frogs

Brilliant Publications – www.brilliantpublications.co.uk
A – Z Maths Games by Karen M. Breitbart

Frog graph

10			
9			
8			
7			
6			
5			
4			
3			
2			
1			

Ghost puzzle

Objective:
Children will practise matching dots to numbers.

Materials:
Ghost (page 37); 1die; scissors; resealable bag.

How to make the game:
♦   Duplicate the ghost, laminate it, cut it out and cut it apart along the puzzle lines.
♦   Store all game pieces and the die in the resealable bag.

How to play the game:
♦   Take the pieces out of the bag and put the puzzle together.
♦   Roll the die and take away the piece of the puzzle that has the same number on it as the number of dots on the die.
♦   Continue to roll and take away pieces until all parts of the ghost are gone. (If you roll a number you've already removed, roll again.)
♦   When you have finished, place all game pieces back in the bag.

Note:
This game can also be played with a spinner. Each section of the spinner should have a different number of dots on it.

Book link:
Gus Loved His Happy Home by Jane Thayer, illustrated by Seymour Fleishman (Linnet Books, 1989). Gus the ghost neglects his house-cleaning chores while Mr Fizzle is on holiday.

Ghost

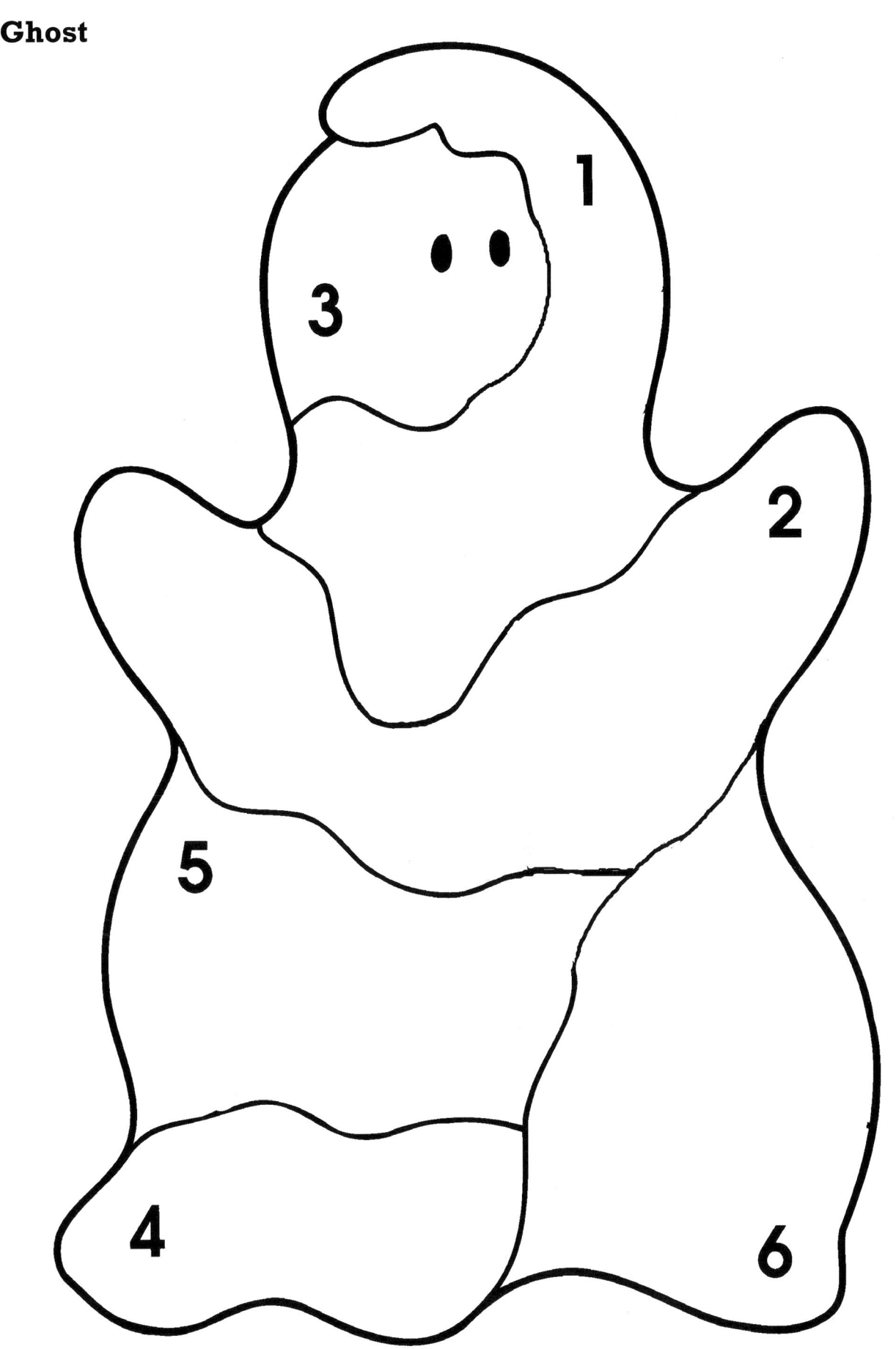

Going for gold

Objective:
Children will solve simple addition problems using counters.

Materials:
Gold coins (page 39); Treasure chests (page 40); crayons or markers (including gold); scissors; counters; resealable bag.

How to make the game:
♦ Duplicate the gold coins, colour gold, laminate and cut out.
♦ Duplicate the treasure chests, colour, laminate and cut out.
♦ Store all game pieces and counters in a resealable bag.

How to play the game:
♦ Spread the game pieces face up on a table. Line the treasure chests in a row and keep the gold coins.
♦ Solve the addition problem on each gold coin. (Use counters if you need help.)
♦ Match each gold coin with the treasure chest that has the correct answer.
♦ When you have finished, ask your teacher to check your work. Then place all game pieces back in the bag.

Book link:
Lucky O'Leprechaun by Jana Dillon (Pelican Publishing Company, 1998). Find out if Meg and Sean manage to catch the leprechaun who lives in their great-aunt's garden.

Brilliant Publications – www.brilliantpublications.co.uk
A – Z Maths Games by Karen M. Breitbart

Gold coins

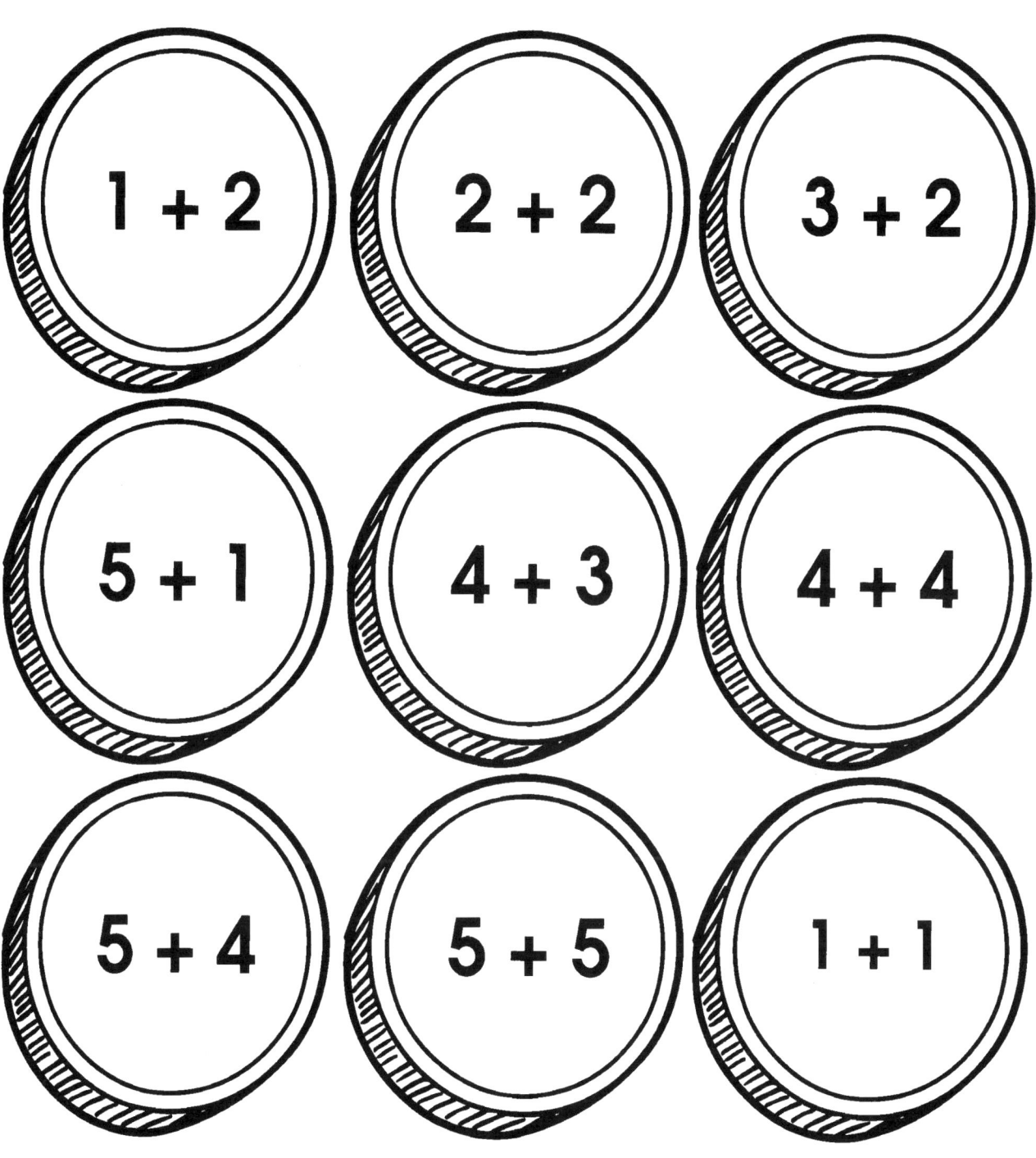

Treasure chests

Brilliant Publications – www.brilliantpublications.co.uk
A – Z Maths Games by Karen M. Breitbart

Halloween maths

Objective:
Children will practise counting and graphing.

Materials:
Halloween things (page 42); Halloween graph (page 43); crayons or markers; scissors; large envelope; resealable bag.

How to make the game:
♦ Duplicate three copies of the Halloween things, colour, laminat, and cut out.
♦ Duplicate one copy of the Halloween graph per child.
♦ Store the Halloween graphs in a large envelope.
♦ Store the things and several crayons in a resealable bag.

How to play the game:
♦ Take one Halloween graph out of the envelope.
♦ Spread the Halloween things face up on a table.
♦ Sort the Halloween things by item.
♦ Use a different colour crayon to mark the number of each item in the correct column on the graph.
♦ When you have finished, give your graph to the teacher, and place the game pieces and the crayons back in the bag.

Option:
Bind the Halloween graphs in a classroom graph book.

Book link:
Ed Emberley's Halloween Drawing Book by Ed Emberley (Little, Brown, 1995). Children will enjoy following these creative drawing tips.

Halloween things

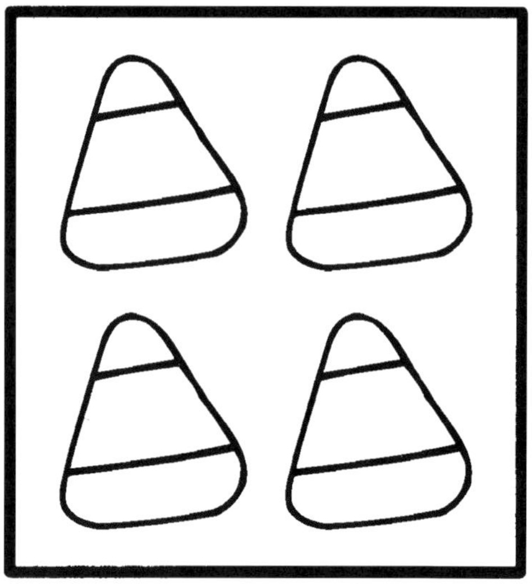

Brilliant Publications – www.brilliantpublications.co.uk
A – Z Maths Games by Karen M. Breitbart

Halloween graph

15					
14					
13					
12					
11					
10					
9					
8					
7					
6					
5					
4					
3					
2					
1					
	bat	candy corn	cat	pumpkin	ghost

Hamburger game

Objective:
Children will match numerals, number words and sets.

Materials:
Hamburger patterns (page 45); sugar paper (red, tan and brown); scissors; permanent marker; resealable bag.

How to make the game:
- Duplicate the hamburger patterns to use as templates.
- Use the templates to cut out ten red tomatoes, ten tan hamburger buns, and ten brown hamburgers.
- Write the numbers from 1 to 10 on the hamburger buns. Write the words 'one' to 'ten'on the hamburgers. Draw sets of dots from one to ten on the tomatoes.
- Laminate all game pieces and cut them out.
- Store the game pieces in a resealable bag.

How to play the game:
- Take the game pieces out of the bag and place them with the numbers, dots, or words faceup on a table.
- Build hamburgers by correctly matching the buns, tomatoes and hamburgers. For example, match the tomato with one dot with the hamburger that has the word 'one' with the bun that has the number 1.
- When you have built ten hamburgers, have your teacher to check your work. Then place all game pieces back in the bag.

Book link:
Cloudy with a Chance of Meatballs by Judi Barrett (Atheneum, 1982). In the town of Chewandswallow, meals come down from the sky.

Hamburger patterns

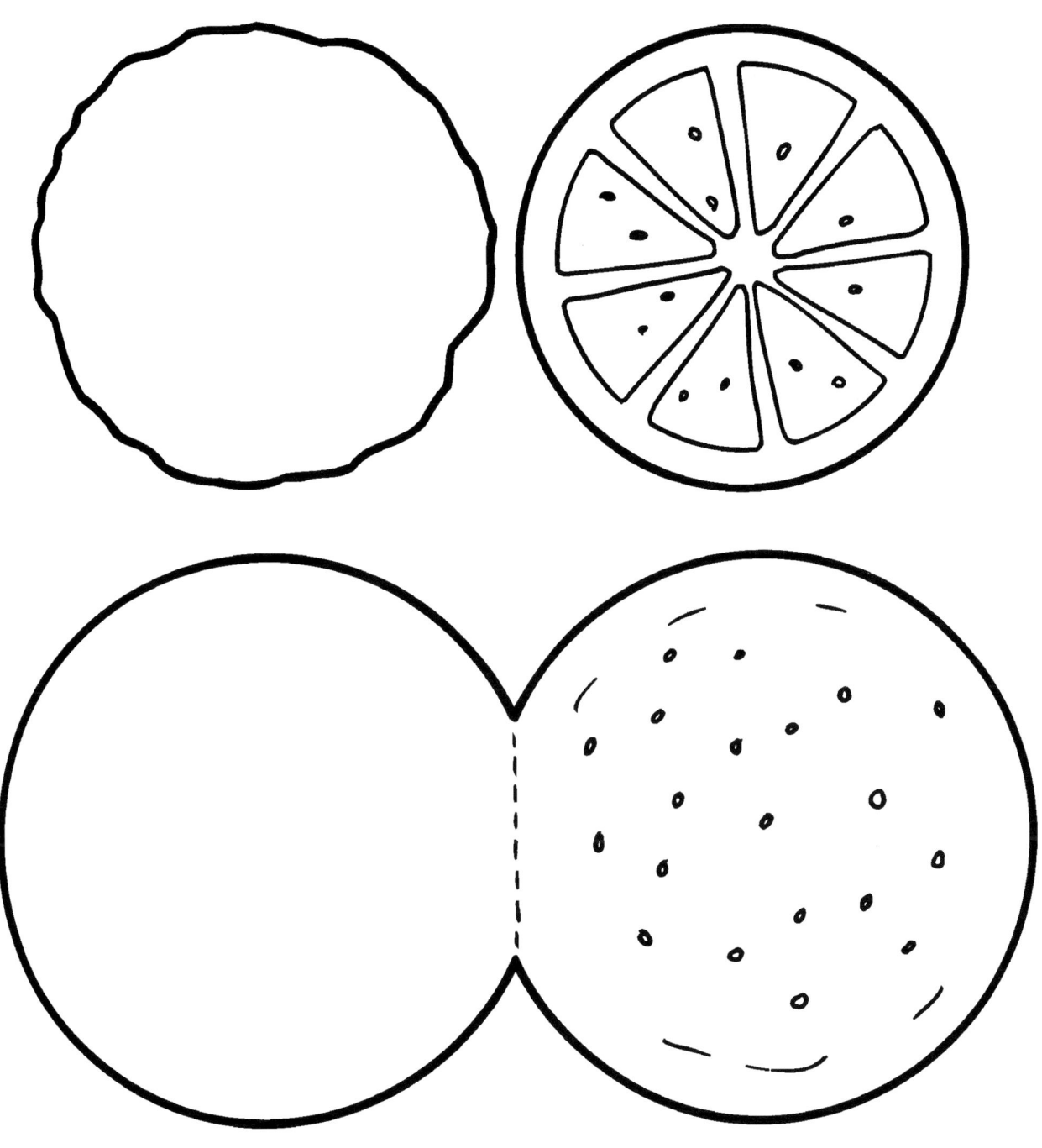

Hot dog game

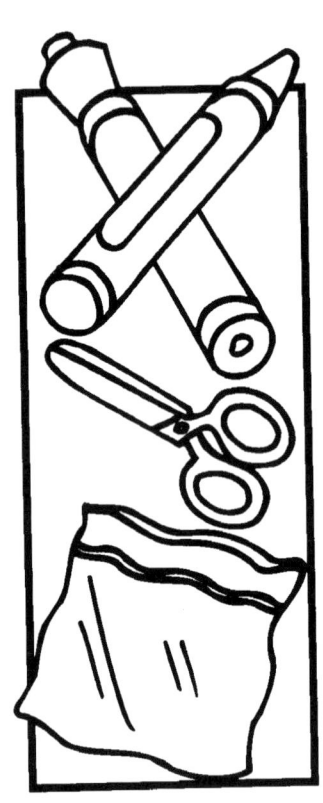

Objective:
Children will match dots to the correct numerals.

Materials:
Hot dogs and buns (page 47); crayons or markers; scissors; resealable bag.

How to make the game:
- Duplicate the hot dogs and buns, colour, laminate and cut out.
- Store the game pieces in a resealable bag.

How to play the game:
- Spread the game pieces face up on a table, and look at the number on each hot dog and the dots on each bun.
- Match the correct hot dog to each bun.
- When you have finished, ask your teacher to check your work. Then place all game pieces back in the bag.

Book link:
Hotter Than a Hot Dog by Stephanie Calmenson (Little, Brown, 1994). A little girl and her grandmother escape the city on a hot summer day by going to the beach.

Hot dogs and buns

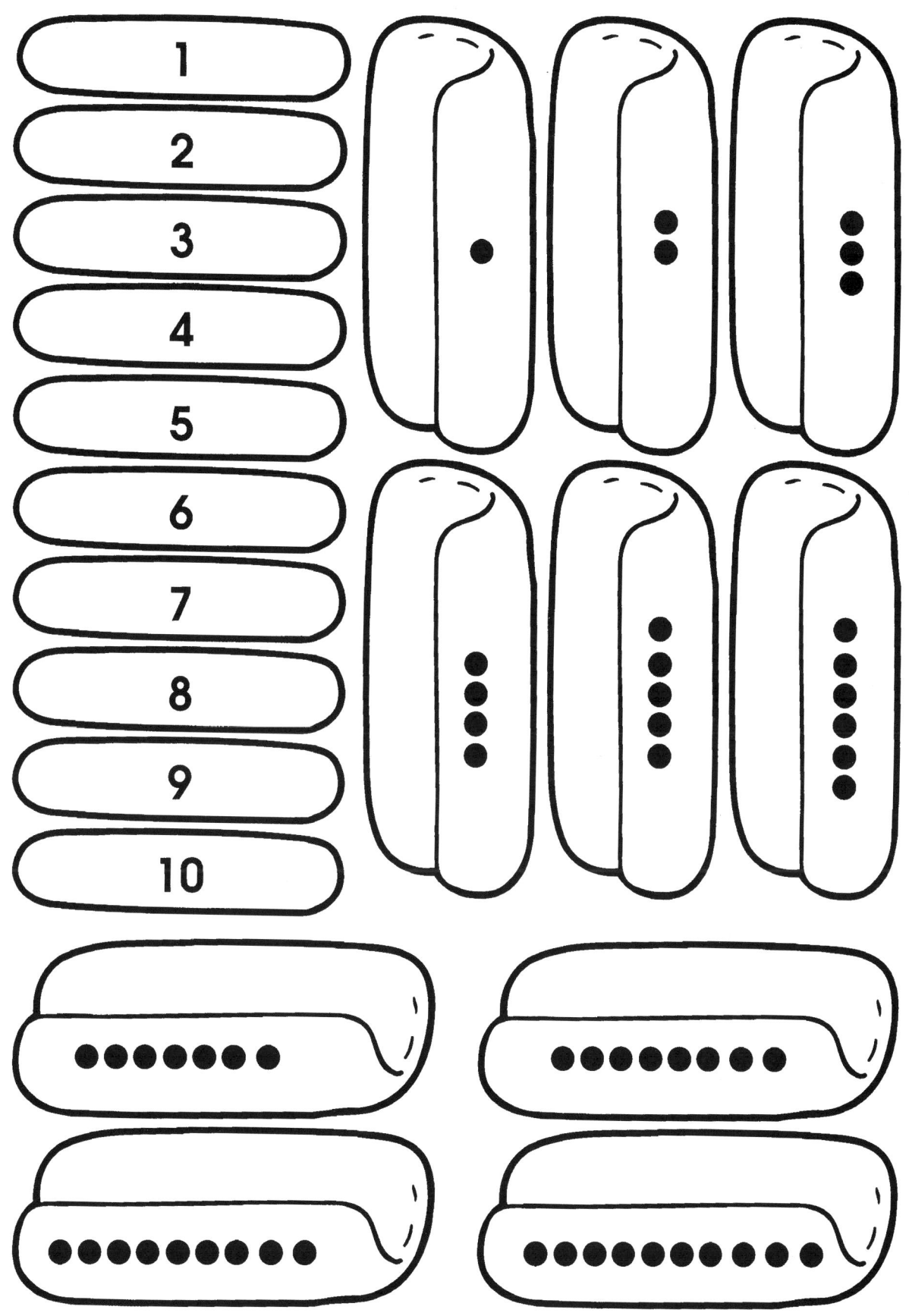

Ice cream match

Objective:
Children will solve simple addition problems using counters.

Materials:
Ice cream cones (page 49); Ice cream scoops (page 50); crayons or markers; scissors; counters; resealable bag.

How to make the game:
♦ Duplicate the ice cream cones and ice cream scoops, colour, laminate and cut out.
♦ Store all game pieces and counters in the resealable bag.

How to play the game:
♦ Take the game pieces out of the bag and spread them face up on a table.
♦ Solve the addition problem on each ice cream scoop. (Use counters to help answer the problems.)
♦ Match the ice cream cone that has the correct answer to each ice cream scoop.
♦ When you have finished, ask your teacher to check your work. Then place all game pieces back in the bag.

Option:
Write the answer on the back of each ice cream scoop for self-checking.

Book link:
Ice-cream Bear by Jez Alborough (Walker Books, 1999). Lazy Bear loves to dream of his favourite thing – ice cream! But one day when he shuts his eyes, Bear gets a shivery surprise!

Ice cream cones

Ice cream scoops

Brilliant Publications – www.brilliantpublications.co.uk
A – Z Maths Games by Karen M. Breitbart

Jaguars' spots

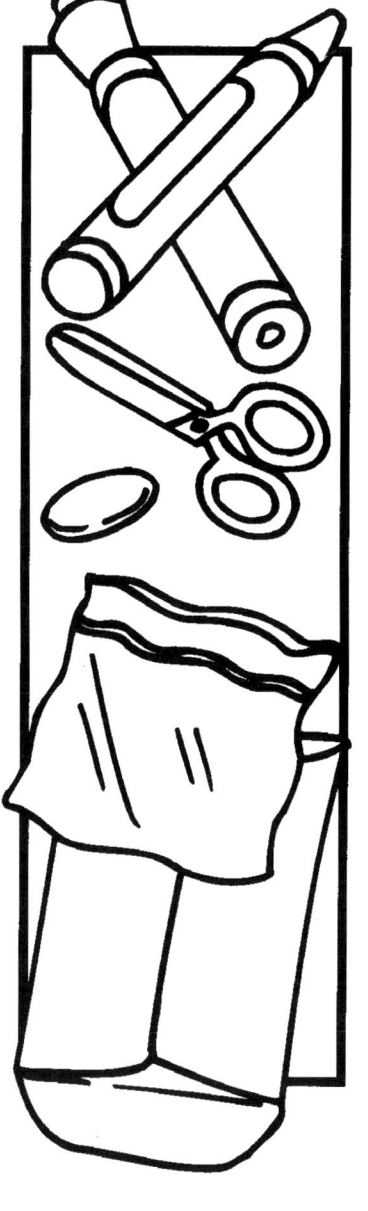

Objective:
Children will practise matching sets to numbers.

Materials:
Jaguars (page 52); crayon; scissors; counters; large envelope; resealable bag.

How to make the game:
♦ Duplicate one copy of the jaguars per child.
♦ Store the jaguars in the large envelope
♦ Store the counters and the crayon in the resealable bag.

How to play the game:
♦ Take one sheet of jaguars out of the envelope.
♦ Count the number of spots on each jaguar. (Circle the spots as you count them to help you keep track.)
♦ Place the matching number of counters on top of each jaguar.
♦ When you have finished, give your teacher your jaguars. Then place the counters and the crayon back in the bag.

Option:
Duplicate the jaguars, colour, laminate and cut out. Have the children match the number of counters to each jaguar's spots.

Book link:
Jaguars in the Rain Forest by Joanne Ryder, illustrated by Michael Rothman (Morrow Junior Books, 1996).

Jaguars

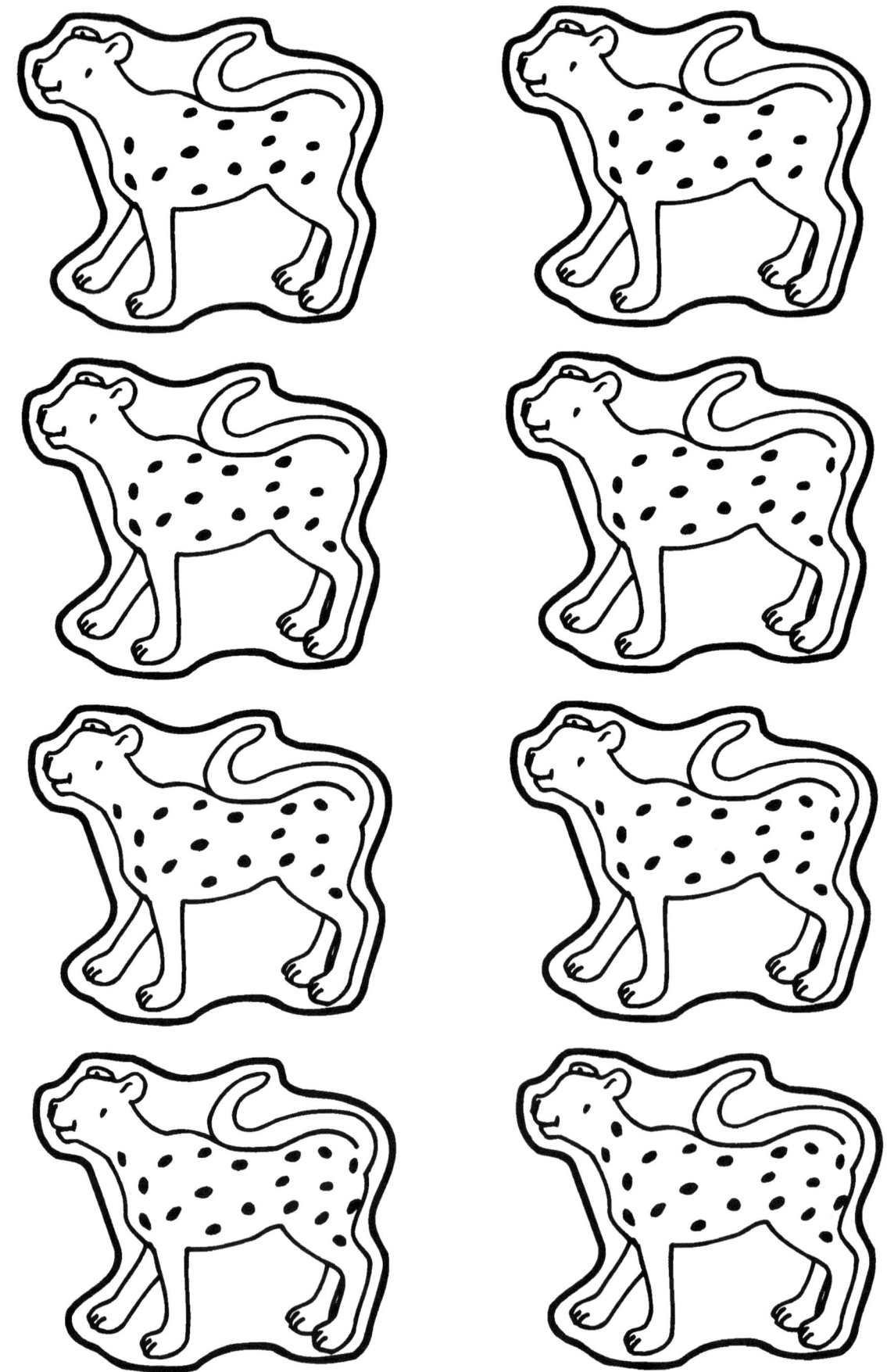

This page may be photocopied by the purchasing institution only.

Brilliant Publications – www.brilliantpublications.co.uk
A – Z Maths Games by Karen M. Breitbart

Jar lids

Objective:
Children will practise estimating, counting and ordering.

Materials:
Estimation sheet (page 54); 5 different-sized jar lids; permanent marker; crayon; counters (beans, beads or buttons); large envelope; resealable bag.

How to make the game:
♦ Duplicate one copy of the estimation sheet per child.
♦ Store the estimation sheets in the large envelope.
♦ Use the permanent marker to number the inside of each jar lid (from 1 to 5).
♦ Store the jar lids, crayon and counters in the resealable bag.

How to play the game:
♦ Take one estimation sheet out of the envelope.
♦ Take the crayon and jar lids out of the bag.
♦ Guess how many counters the first jar lid might hold.
♦ Write the number in the first square of the 'guess' column.
♦ Scoop up the counters with the jar lid and then count how many are in the scoop.
♦ Write this number in the first square of the 'answer' column.
♦ Continue to guess, count and record for each jar lid.
♦ When you have finished, place all the game pieces back in the bag. Give your estimation sheet to the teacher.

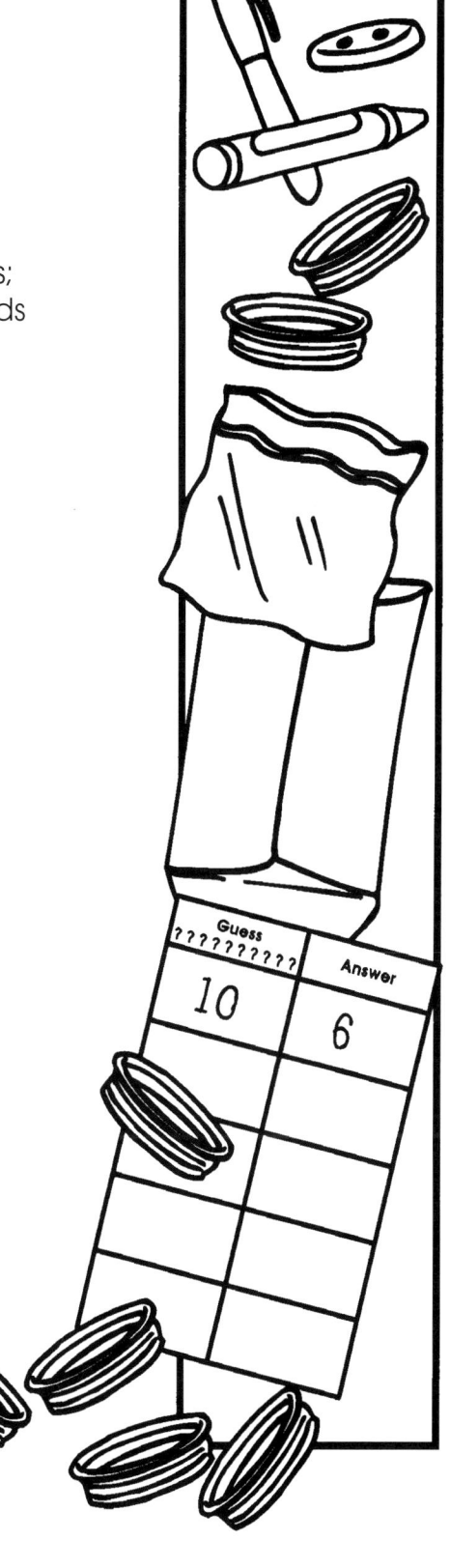

Book link:
Uno, Dos, Tres: One, Two Three by Pat Mora, illustrated by Barbara Lavalee (Clarion, 2000). Rhyming text presents numbers in English and Spanish.

Estimation sheet

Guess	Answer

Brilliant Publications – www.brilliantpublications.co.uk
A – Z Maths Games by Karen M. Breitbart

Kings' crowns

Objective:
Children will practise making sets to match numerals.

Materials:
Crown (page 56); Jewels (page 57); crayons or markers; scissors; large envelope; resealable bag.

How to make the game:
- Make ten copies of the crown and three copies of the jewels.
- Number the crowns from 1 to 10.
- Colour, laminate and cut out the crowns and jewels.
- Store the crowns in the large envelope and the jewels in the resealable bag.

How to play the game:
- Take the crowns out of the envelope and line them up in a row, from 1 to 10.
- Take the jewels out of the bag and use them to make number sets for each crown. For example, place one jewel on the first crown, two on the second and so on.
- When you have finished, ask your teacher to check your work. Then put the crowns in the envelope and the jewels in the bag.

Options:
- Decorate the crowns and jewels using glitter and glue.
- Let children make sets of beads to place on the crowns.

Book link:
King Bidgood's in the Bathtub by Audrey Wood, illustrated by Don Wood (Child's Play (International) 5, 1989).
A fun loving king refused to get out of the bath.

Crown

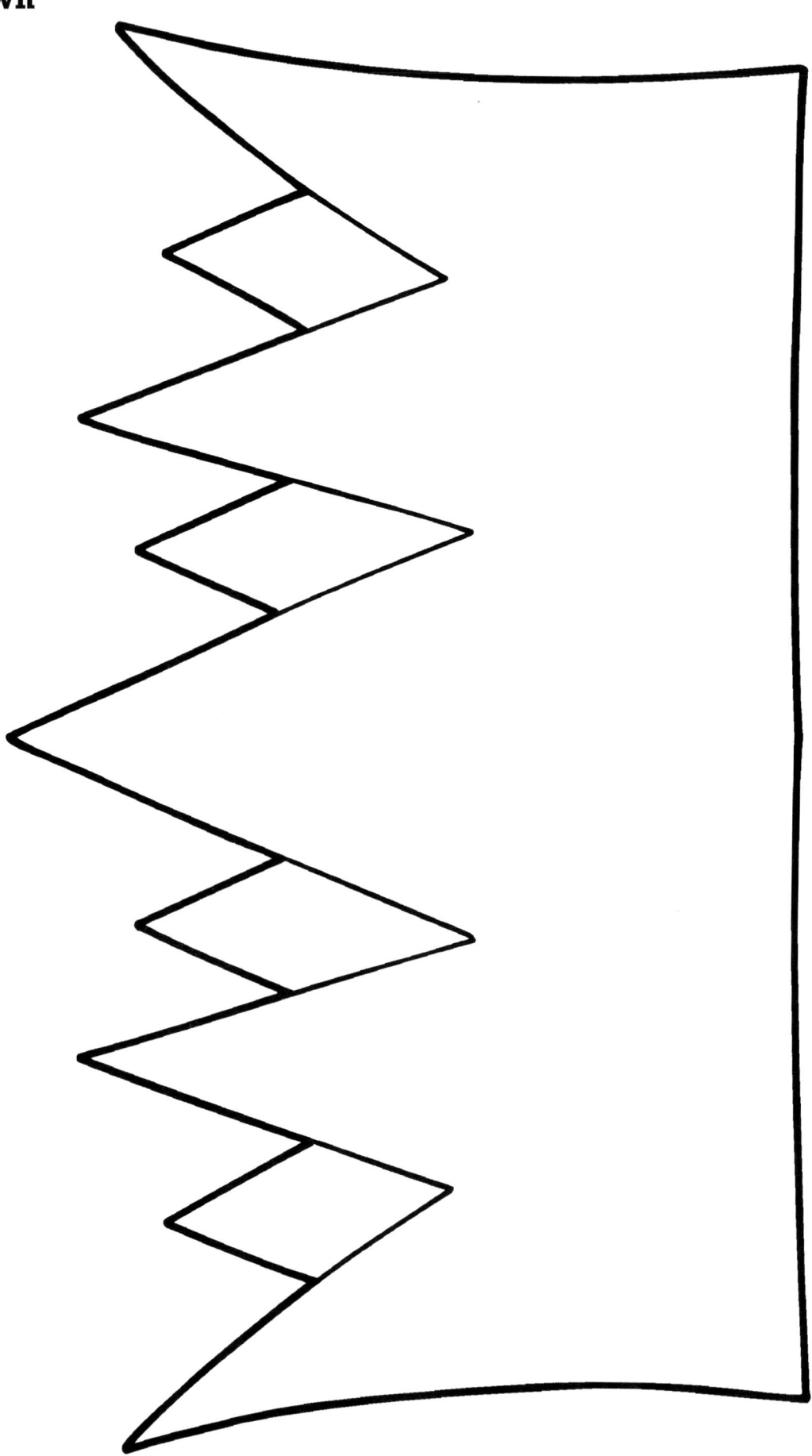

Brilliant Publications – www.brilliantpublications.co.uk
A – Z Maths Games by Karen M. Breitbart

Jewels

Kite game

Objective:

Children will make sets of numbers from 1 to 10.

Materials:

Kite (page 59); 55 clothes-pegs; wool or ribbon; scissors; hole punch; crayons or markers; large envelope; resealable bag.

How to make the game:

♦ Make ten copies of the kite and number them from 1 to 10.
♦ Colour the kites, laminate and cut out.
♦ Punch a hole in the bottom of each kite and attach a piece of wool or ribbon to represent the tail.
♦ Store the kites in the large envelope and the clothes-pegs in the resealable bag.

How to play the game:

♦ Take the kites out of the envelope, and line them up from 1 to 10.
♦ Attach the matching number of clothes-pegs to each kite tail.
♦ When you have finished, ask your teacher to check your work. Then place the clothes-pegs in the bag and the kites in the envelope.

Option:

Use different-coloured plastic clothes-pegs and colour the kites accordingly. (Colour kite 1 blue, and place one blue clothes-peg in the bag. Colour kite 2 red and place two red clothes-pegs in the bag, and so on.)

Book link

Alone in the Woods by Ian Beck (Scholastic David Fickling Books, 2000). Lily's kite carries Teddy up into the sky. But how does he get back?

Brilliant Publications – www.brilliantpublications.co.uk
A – Z Maths Games by Karen M. Breitbart

Kite

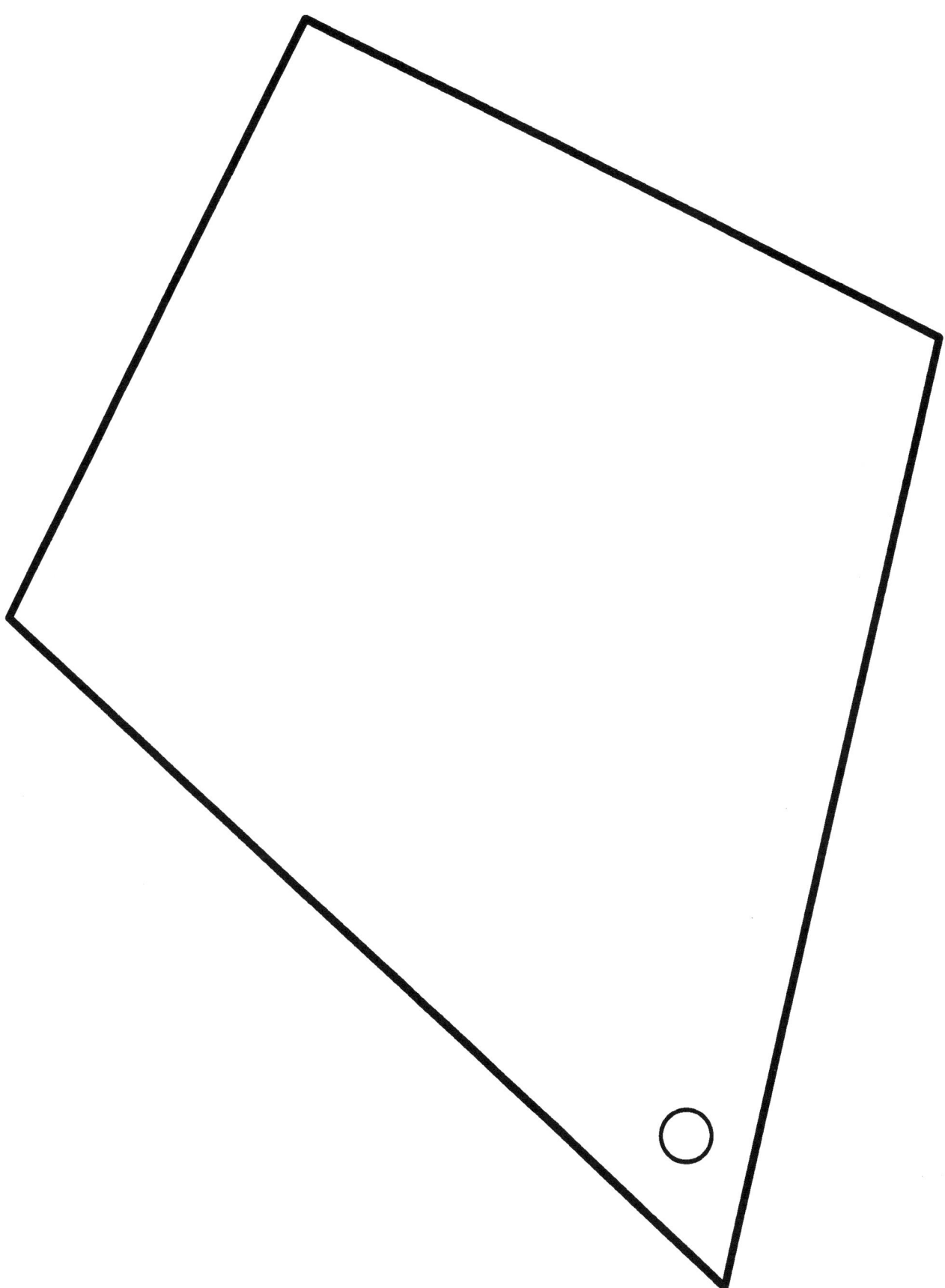

Ladybird game

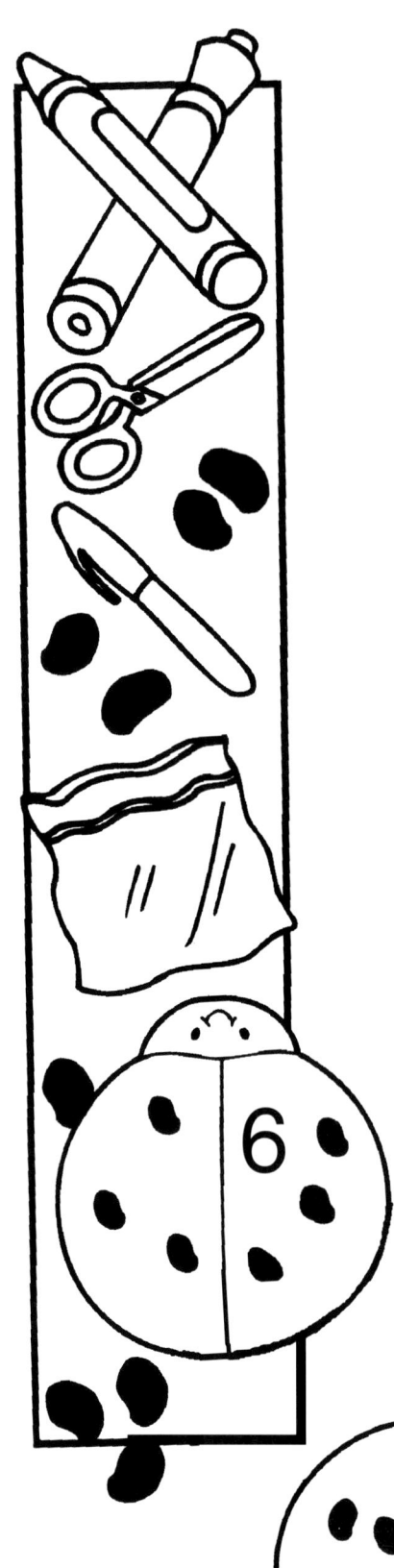

Objective:
Children will make sets of the numbers 1 to 15.

Materials:
Ladybirds (page 61); permanent marker; crayons or markers; scissors; black beans (or other black counters); resealable bag.

How to make the game:
♦ Duplicate five copies of the ladybirds and number the ladybirds' shells from 1 to 15.
♦ Colour the ladybirds, laminate and cut out.
♦ Store the ladybirds and the black beans in a resealable bag.

How to play the game:
♦ Spread the ladybirds face up on the floor.
♦ Look at the number on each ladybird.
♦ Use the black beans to put the correct number of spots on the wings of each ladybird.
♦ When you have finished, ask you teacher to check your work. Then place the ladybirds and the counters back in the bag.

Book link:
Are you a Ladybird? Judy Allen, Tudor Humpries (Kingfisher Books, 2000). The text and colourful artwork bring this familiar small creature to life.

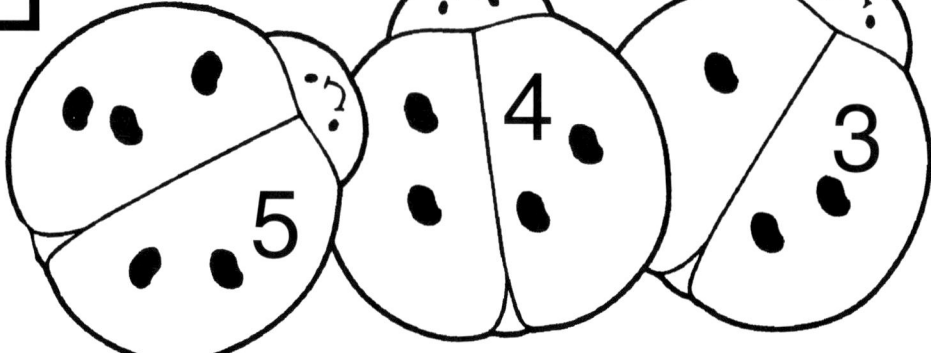

Ladybirds

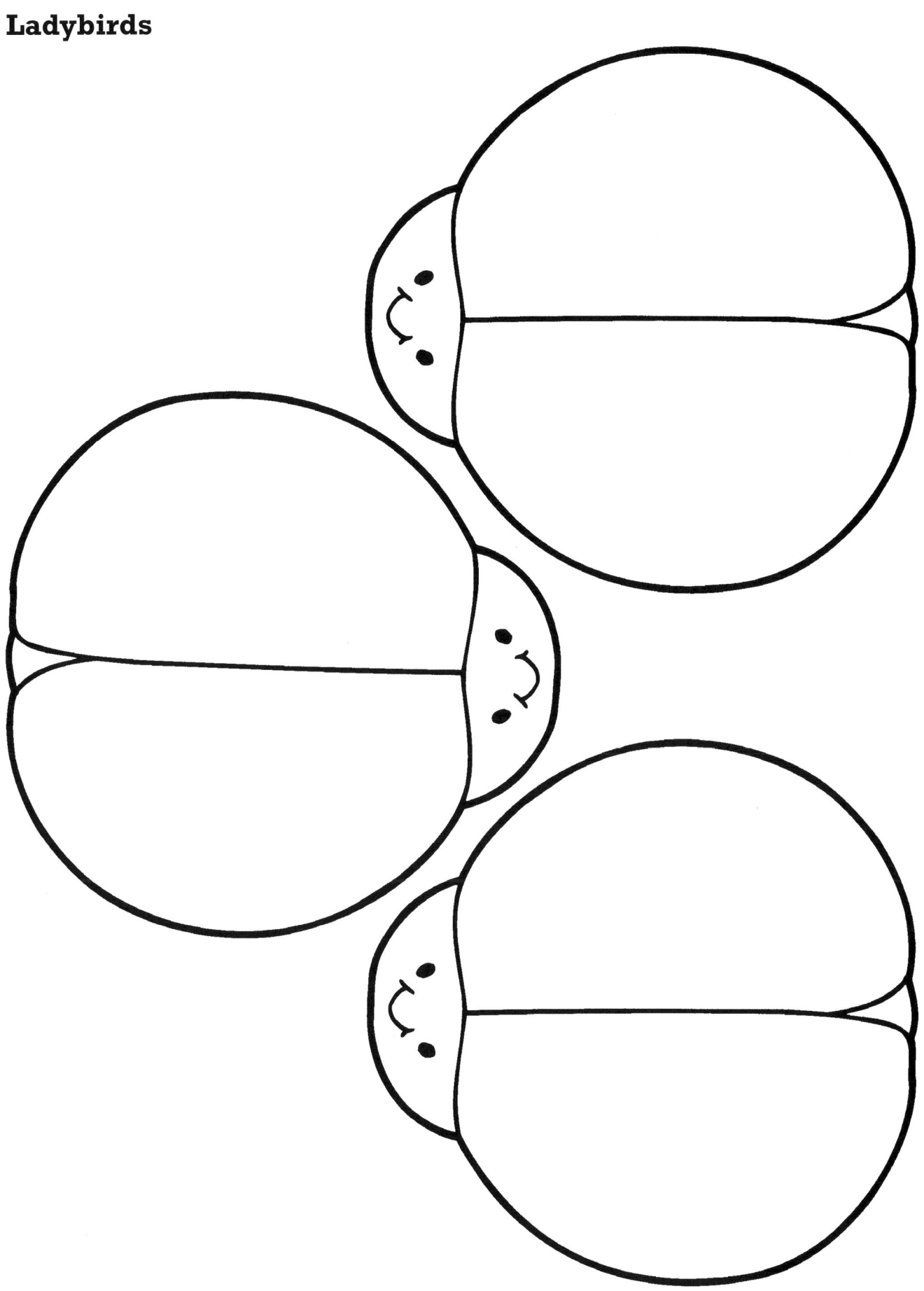

Measuring

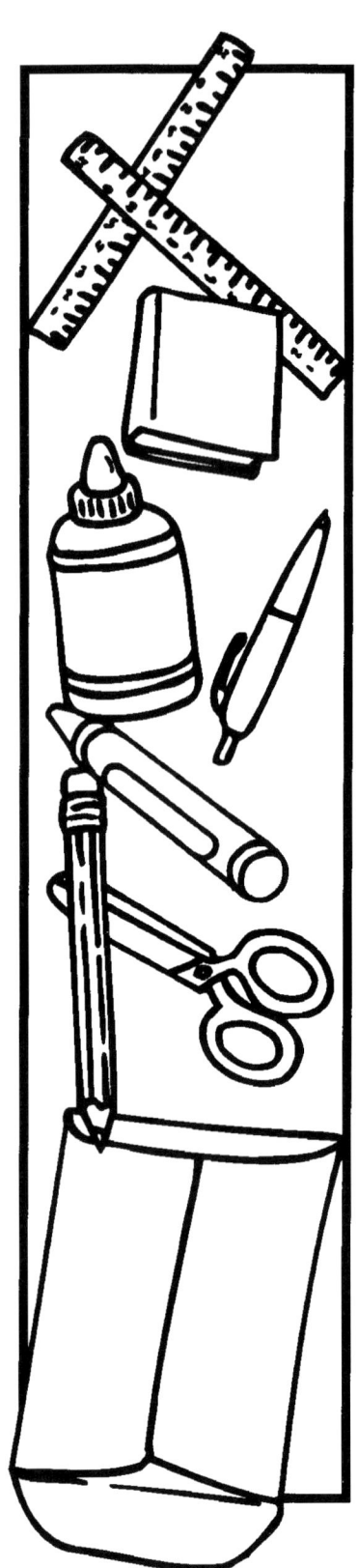

Objective:
Children will practise estimating and measuring.

Materials:
Measuring sheet (page 63); 2 rulers; large envelope; book; child's scissors; pen; pencil; crayon; bottle of glue.

How to make the game:
♦ Duplicate one copy of the measuring sheet per child.
♦ Store the measuring sheets in the large envelope.
♦ Place the ruler and the items to be measured (second ruler, book, child's scissors, pen, pencil, crayon, and bottle of glue) in the maths basket.

How to play the game:
♦ Take a measuring sheet out of the large envelope.
♦ Look at the items to be measured on the sheet, and guess how many centimetres long each item is. Write this number in the first box on the measuring sheet.
♦ Use the ruler to measure each item and write the length in the second box 'Measurement' on the measuring sheet.
♦ Compare your guesses with the correct answers.

Book link:
Inch by Inch: The Garden Song by David Mallett (HarperCollins, 1997). A child grows a garden with the help of the rain and the earth.

Measuring sheet

Name _____

Items	Guess	Measurement
ruler		
book		
scissors		
pen		
pencil		
crayon		
glue		

More or less

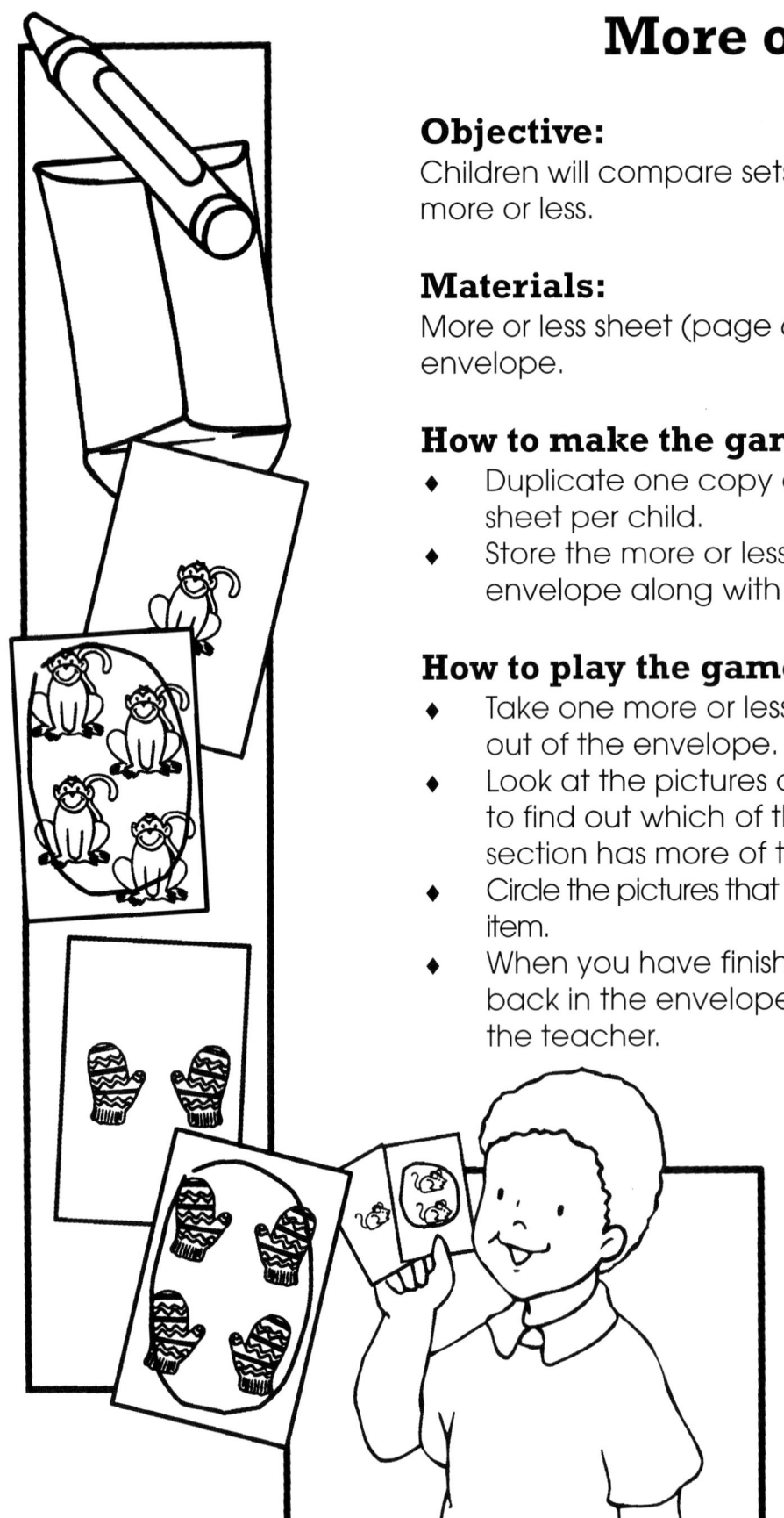

Objective:
Children will compare sets to find out which has more or less.

Materials:
More or less sheet (page 65); crayon; large envelope.

How to make the game:
♦ Duplicate one copy of the more or less sheet per child.
♦ Store the more or less sheets in the large envelope along with the crayon.

How to play the game:
♦ Take one more or less sheet and the crayon out of the envelope.
♦ Look at the pictures on the page and count to find out which of the two pictures in each section has more of the item.
♦ Circle the pictures that have the most of each item.
♦ When you have finished, place the crayon back in the envelope and give your sheet to the teacher.

Options:
Ask the children to choose two handfuls of counters and compare the sets, counting to find which handful has more than the other. Ask the children use counters to make sets for the items on the more or less sheet.

Book link:
How Many? by Debbie MacKinnon (Frances Lincoln, 1997).

More or less sheet

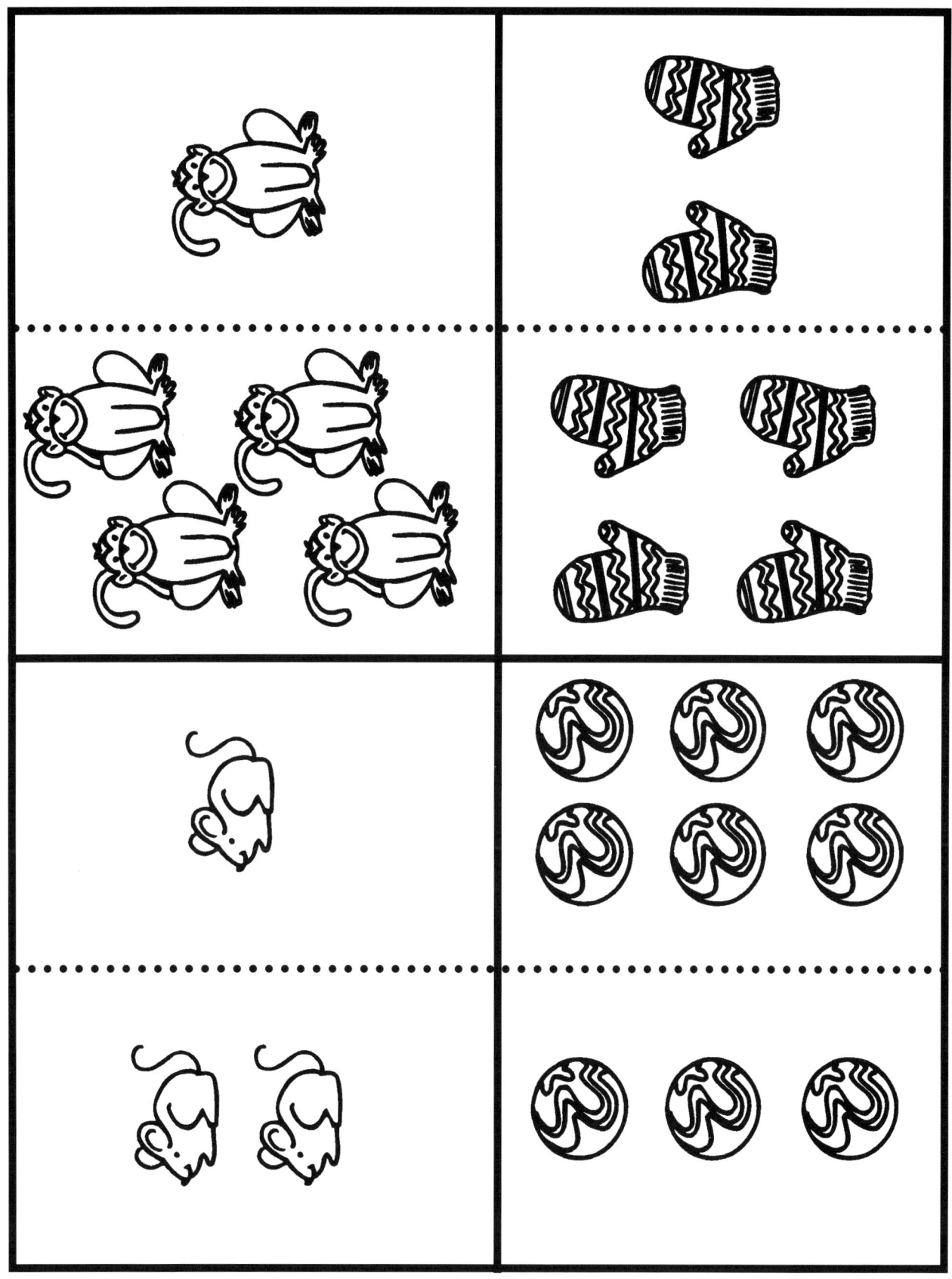

Musical maths

Objective:
Children will practise repetitive rhythm patterns.

Materials:
Instruments (maracas, rattle, triangle, bells, and so on); shoe box or other small box; 3 index cards; marker.

How to make the game.
♦ Place the instruments in the box.
♦ On the index cards draw different patterns, such as a long line followed by two dots followed by another line, and so on. (See suggested patterns at the bottom of the page.)
♦ Store the cards in the box with the intruments.

How to play the game:
♦ Take the cards and the instruments out of the box.
♦ Look at the pattern on the first card. The long lines represent a long note, and the dots represent short notes.
♦ Use the musical instruments to play each pattern. For example, if the card has a long line followed by two dots followed by another long line, you would use the instrument to play a long note, two quick notes, and another long note.
♦ Play the pattern on each card.
♦ When you have finished, place the cards and the instruments back in the box.

Suggested patterns:

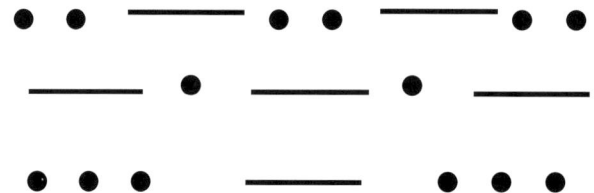

This is a slightly noisy activity.

Number necklace

Objective:
Children will put the numbers 1 to 10 in the correct order.

Materials:
Tubular macaroni (10 pieces per set); permanent marker; wool; scissors; resealable bags.

How to make the game:
♦ Write one number from 1 to 10 on each piece of macaroni. (Make sure each set has one 1, one 2, and so on, through to 10.)
♦ Cut a necklace-length piece of wool for each child.
♦ Store the sets in individual bags.

How to play the game:
♦ Take the wool and the pieces of macaroni out of the bag.
♦ Line the macaroni pieces in order from 1 to 10.
♦ String the pieces of macaroni in order on the piece of wool.
♦ Show the necklace to your teacher when you've finished. Your teacher can help you tie the ends of your necklace together.

Book link:
Ants in Your Pants by Sue Boyle, illustrated by Sue Heap (Scholastic, 1997). Introduces counting in a way that is both funny and memorable.

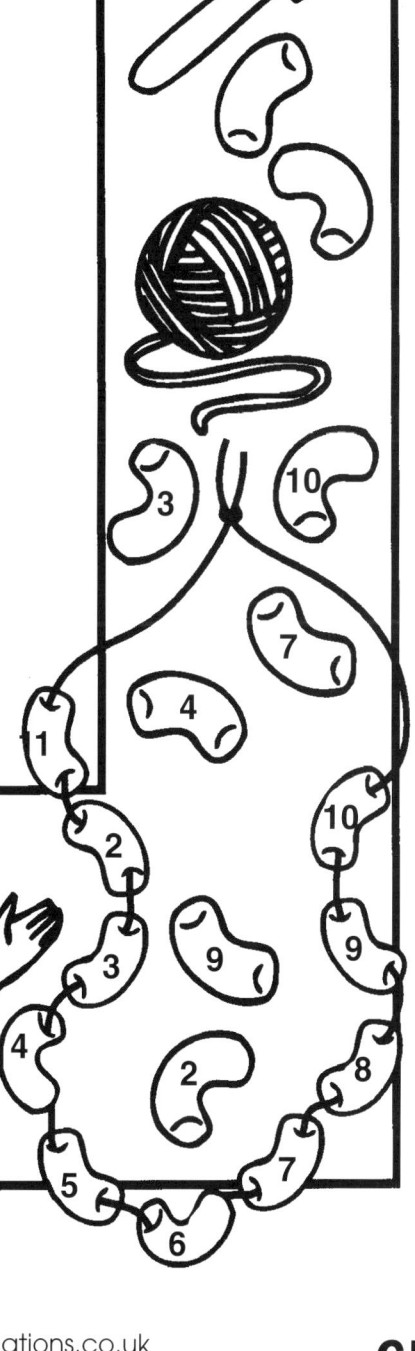

Ollie the Octopus

Objective:
Children will practise making sets of the numbers 1 to 8.

Materials:
Octopus (page 69); loop cereal (or hole reinforcers); glue; large envelope; resealable bag.

How to make the game:
♦ Duplicate a copy of the octopus for each child.
♦ Store the octopi in the large envelope.
♦ Store the cereal in a resealable bag.
♦ Place the bottle of glue in the Maths Centre.

How to play the game:
♦ Take one copy of the octopus out of the envelope.
♦ Look at the number on each of the octopus's legs.
♦ Glue the correct number of cereal pieces to each one of the octopus's legs to represent the numbers.
♦ When you have finished, give your octopus to the teacher and place the rest of the materials in the Maths Centre.

Option:
Post the completed Octopus art on an 'Under the sea' bulletin board. Let the children decorate it using green and blue streamers.

Book link:
Oscar Orange and the Octopus by Stephanie Laslett and Maggie Downer (HarperCollins, 1995).

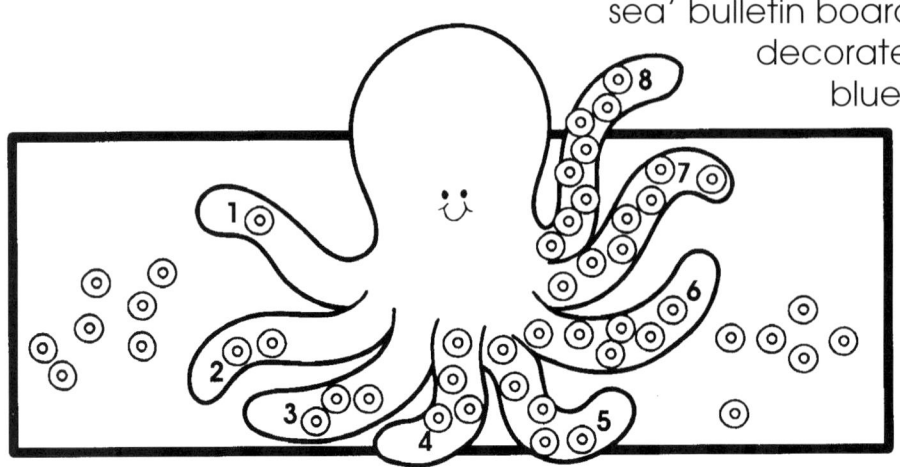

Brilliant Publications – www.brilliantpublications.co.uk
A – Z Maths Games by Karen M. Breitbart

Octopus

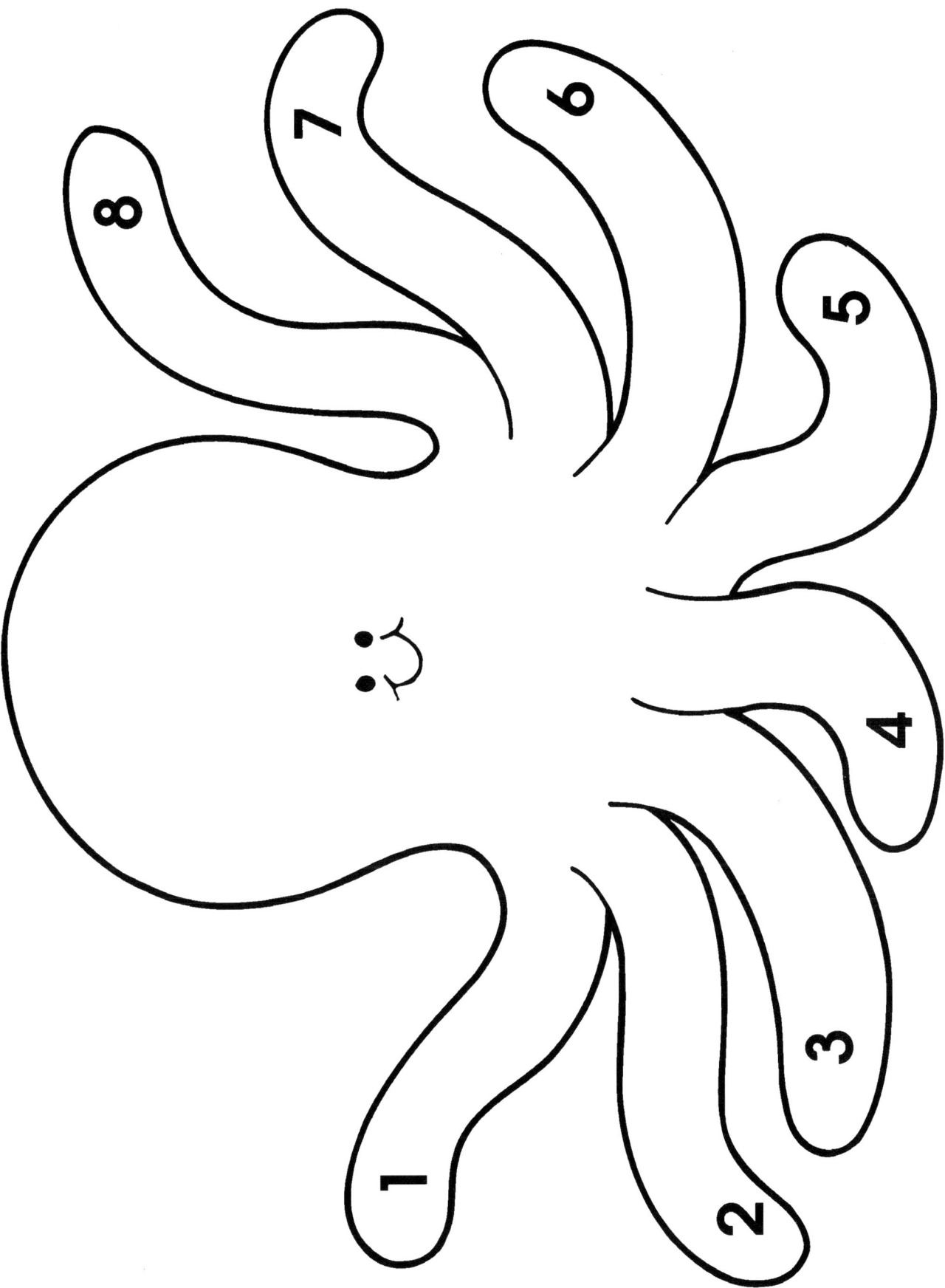

Ostrich eggs

Objective:
Children will be able to put numbers in order from 1 to 20.

Materials:
Ostrich eggs (page 71); crayons or markers; scissors; resealable bag.

How to make the game:
♦ Duplicate 10 copies of the ostrich eggs.
♦ Number the ostrich eggs from 1 to 20.
♦ Laminate the eggs and cut them out.
♦ Store the ostrich eggs in a resealable bag.

How to play the game:
♦ Take the ostrich eggs out of the envelope.
♦ Line the eggs up in order from 1 to 20.
♦ Mix the eggs, and line them up again, this time from 20 to 1.
♦ When you have finished, ask your teacher to check your work. Then place all the ostrich eggs back in the bag.

Options:
♦ Share this fact with your children: ostrich eggs weigh between 1.0 to 1.78 kg.
♦ Provide weights and a scale for them to see how heavy this is.
♦ Bring in a chicken's egg for children to weigh in comparison.

Book link:
Ostrich Big Book by Seddon (Cambridge University Press, 1998).

Brilliant Publications – www.brilliantpublications.co.uk
A – Z Maths Games by Karen M. Breitbart

Ostrich eggs

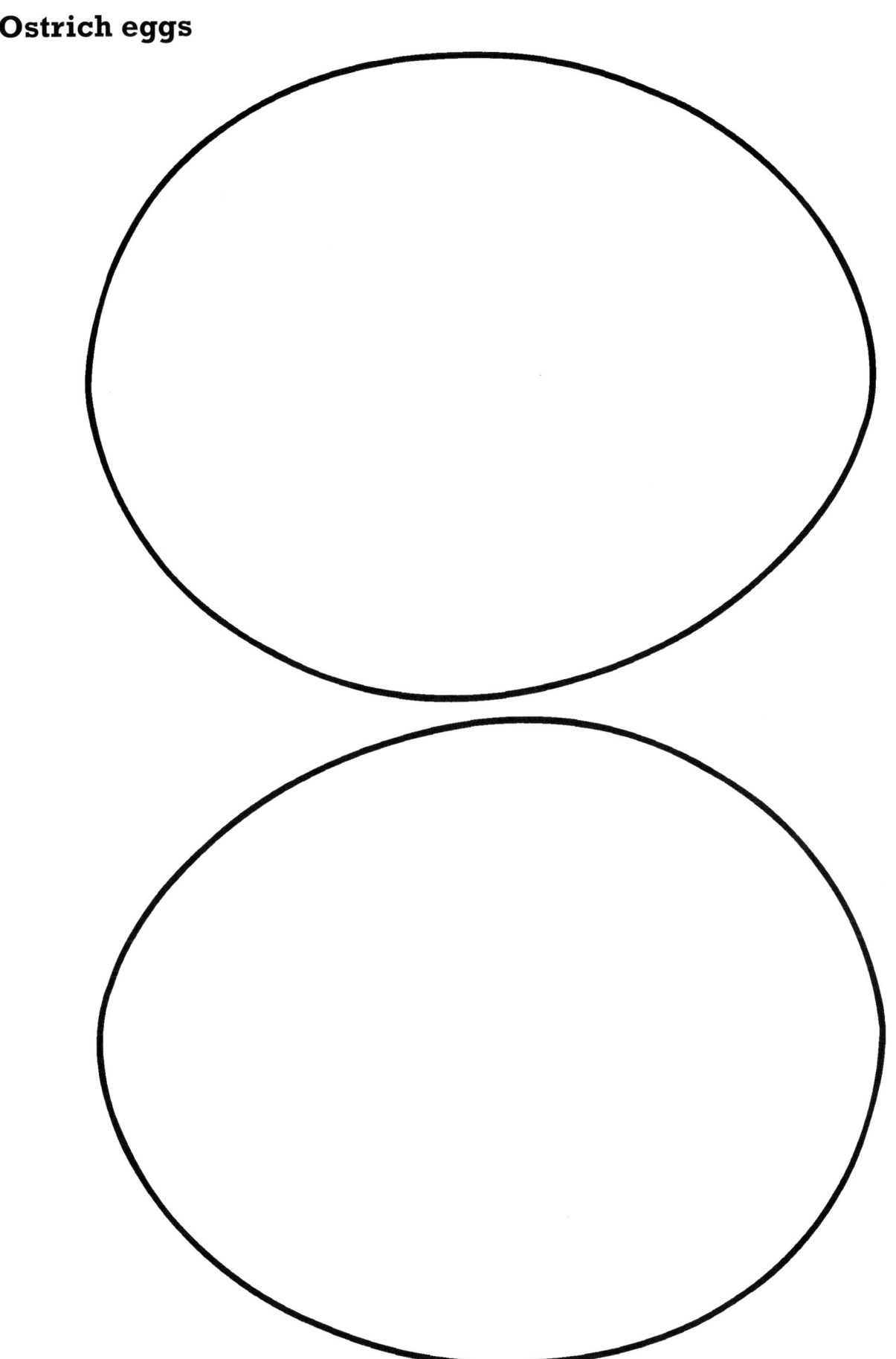

Pudding paint

Objective:
The children will practise writing the numbers 1 to 20.

Materials:
Instant pudding (plus ingredients to make it); paper plates (one per child); index cards; crayons or markers; spoon.

How to make the game:
♦ Write the numbers 1 to 20 on the index cards and laminate them.
♦ Make the instant pudding. If possible, let the children help you.

How to play the game:
♦ Spoon out a small amount of pudding onto a paper plate.
♦ Trace the numbers from 1 to 20 in the pudding.
♦ If the children can't remember what a number looks like, look at the index card.
♦ When you have finished, wash your hands and make sure there is no pudding on the table.

Note:
This activity should be done as a class.

Options:
♦ Make enough pudding to serve it as a snack after the children have drawn numbers.
♦ Use finger paint instead of pudding.

Book link:
Roly-Poly Pudding by Beatrix Potter (Warne, 1989). This is the tale of Samuel Whiskers, or the roly-poly pudding.

Brilliant Publications – www.brilliantpublications.co.uk
A – Z Maths Games by Karen M. Breitbart

Patterns

Objective:
Children will make their own patterns.

Materials:
Coloured loop-style cereal (or coloured macaroni); straws; glue; scissors; index cards; necklace-sized lengths of wool (one per child); 2 resealable bags.

How to make the game:
♦ Cut the straws into 2.5 cm pieces.
♦ Make several sample patterns by gluing the cereal and straw pieces to the index cards.
♦ Store the cereal, straws and wool in one resealable bag.
♦ Store the sample patterns in a separate resealable bag.

How to play the game:
♦ Take one piece of wool out of the bag.
♦ Take one of the index cards (with cereal and straws glue on it) out of the bag or make up your own pattern.
♦ Use the cereal pieces and straw pieces to make a necklace that follows the same pattern as that on the index card. A pattern means that you will put things in order. For example, two pieces of cereal followed by one straw followed by two pieces of cereal.
♦ Show the necklace to your teacher when you've finished. Your teacher can help you tie the ends of the necklace together.
♦ Replace the other materials in the bags.

Book link:
Colour Me Crazy Andrew Donkin, Jeff Cummin (HarperCollins, 1996). Kevin's town is the most boring place on earth, until the spectrums arrive to brighten things up.

Postman

Objective:
Children will practise reading three-digit numbers.

Materials:
Postman's hat (page 75); houses (page 76); letters (page 77); blue sugar paper; glue or tape; crayons or markers; scissors; resealable bag.

How to make the game:
♦ Duplicate the postman's hat onto blue sugar paper, colour, laminate, and cut out. Fold the brim upwards on the dotted line.
♦ Make a sugar paper band and attach it to the postman's hat pattern. (Make sure the band is long enough to fit around a child's head.)
♦ Tape or glue the ends of the band together to make the hat.
♦ Copy the houses and letters, colour, laminate and cut out.
♦ Store all game pieces in a resealable bag.

How to play the game:
♦ Take the postman's hat from the bag and put it on.
♦ Line up the houses in a row and take the letters out of the bag.
♦ Look at the numbers on the houses and the letters.
♦ Deliver the post by matching each letter to the correct house.
♦ When you have finished, ask your teacher to check your work. Then place all game pieces back in the bag.

Book link:
The Jolly Postman or Other People's Letters by Janet and Allan Ahlberg (Viking Children's Books, 1999). This inventive book includes real letters in envelopes.

Postman's hat

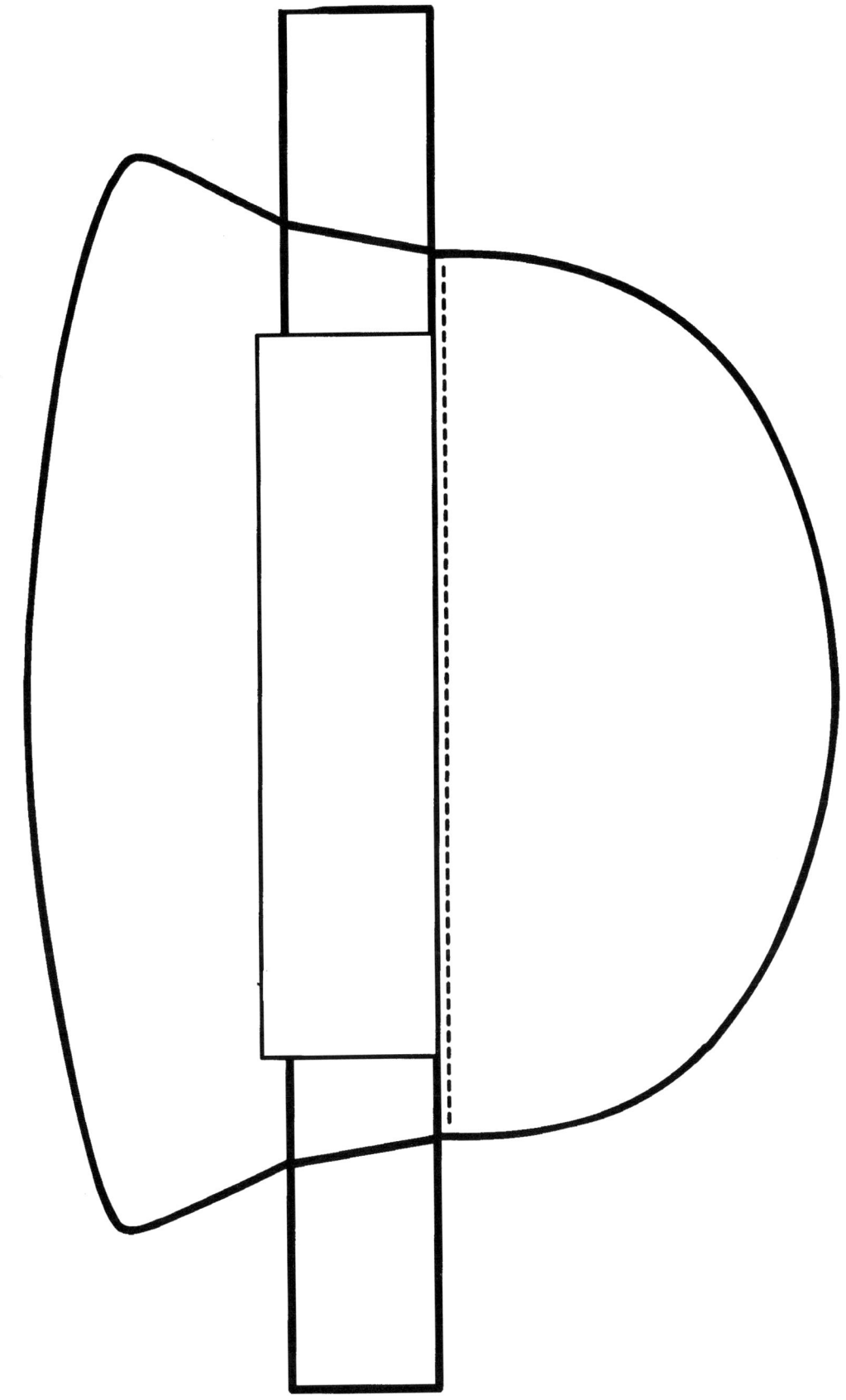

Houses

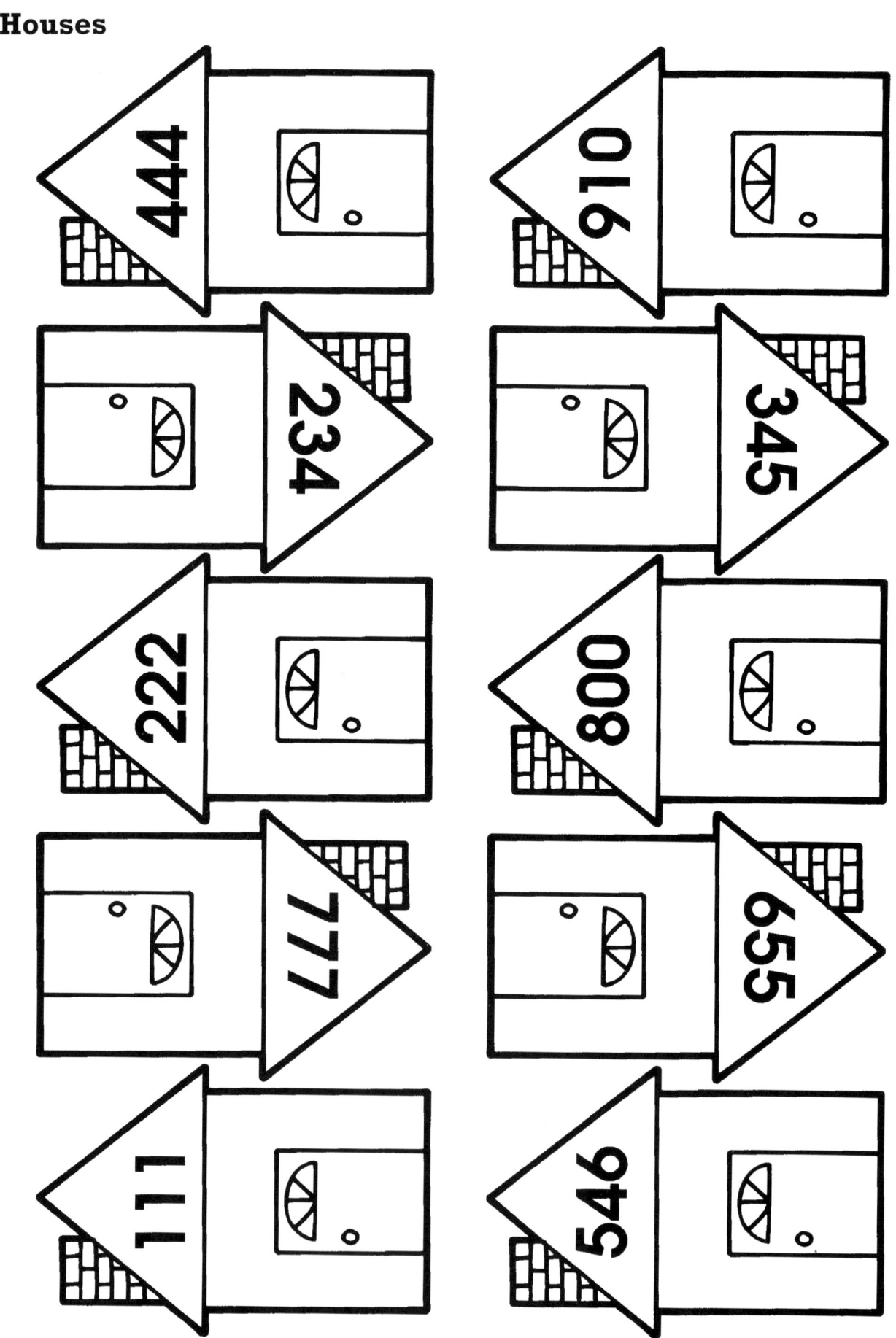

Brilliant Publications – www.brilliantpublications.co.uk
A – Z Maths Games by Karen M. Breitbart

Letters

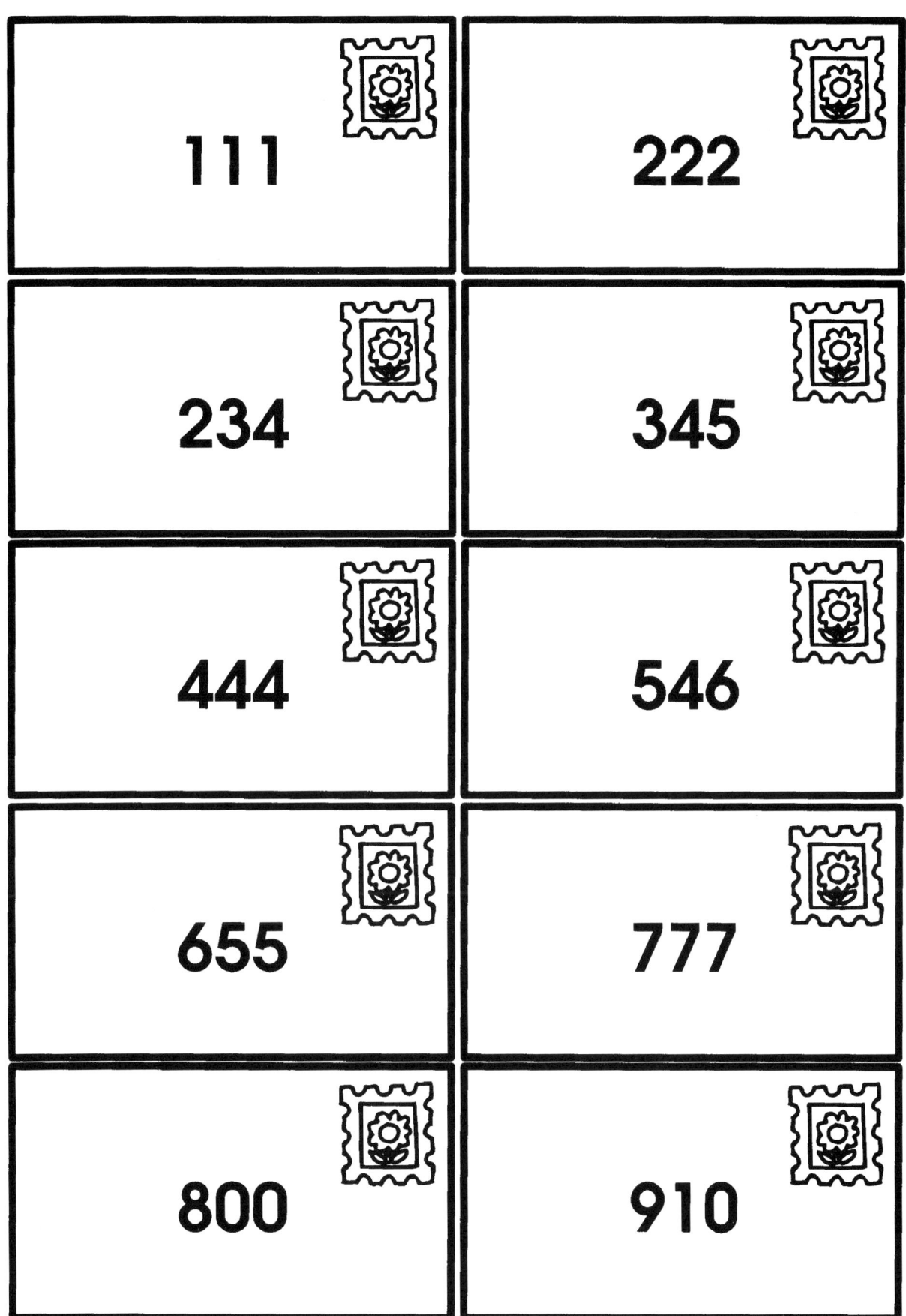

111

222

234

345

444

546

655

777

800

910

Brilliant Publications – www.brilliantpublications.co.uk
A – Z Maths Games by Karen M. Breitbart

Quilted numbers

Objective:
Children will match sets to numerals from 1 to 9.

Materials:
Quilt (page 79); crayons or markers; scissors; large envelope; small counters; resealable bag.

How to make the game:
♦   Duplicate the quilt, colour, laminate and cut out.
♦   Place the counters in a resealable bag.
♦   Store the quilt in a large envelope.

How to play the game:
♦   Take the quilt out of the envelope.
♦   Look at the number on each quilt square.
♦   Place the correct number of counters on top of each quilt square to represent each number.
♦   When you have finished, ask your teacher to check your work. Then place the counters back in the bag and the quilt back in the envelope.

Book link:
Luka's Quilt by Georgia Guback (Greenwillow, 1994). There is a disagreement over what colours Luka's traditional Hawaiian quilt should be.

Brilliant Publications – www.brilliantpublications.co.uk
A – Z Maths Games by Karen M. Breitbart

Quilt

Brilliant Publications – www.brilliantpublications.co.uk
A – Z Maths Games by Karen M. Breitbart

Counting quails

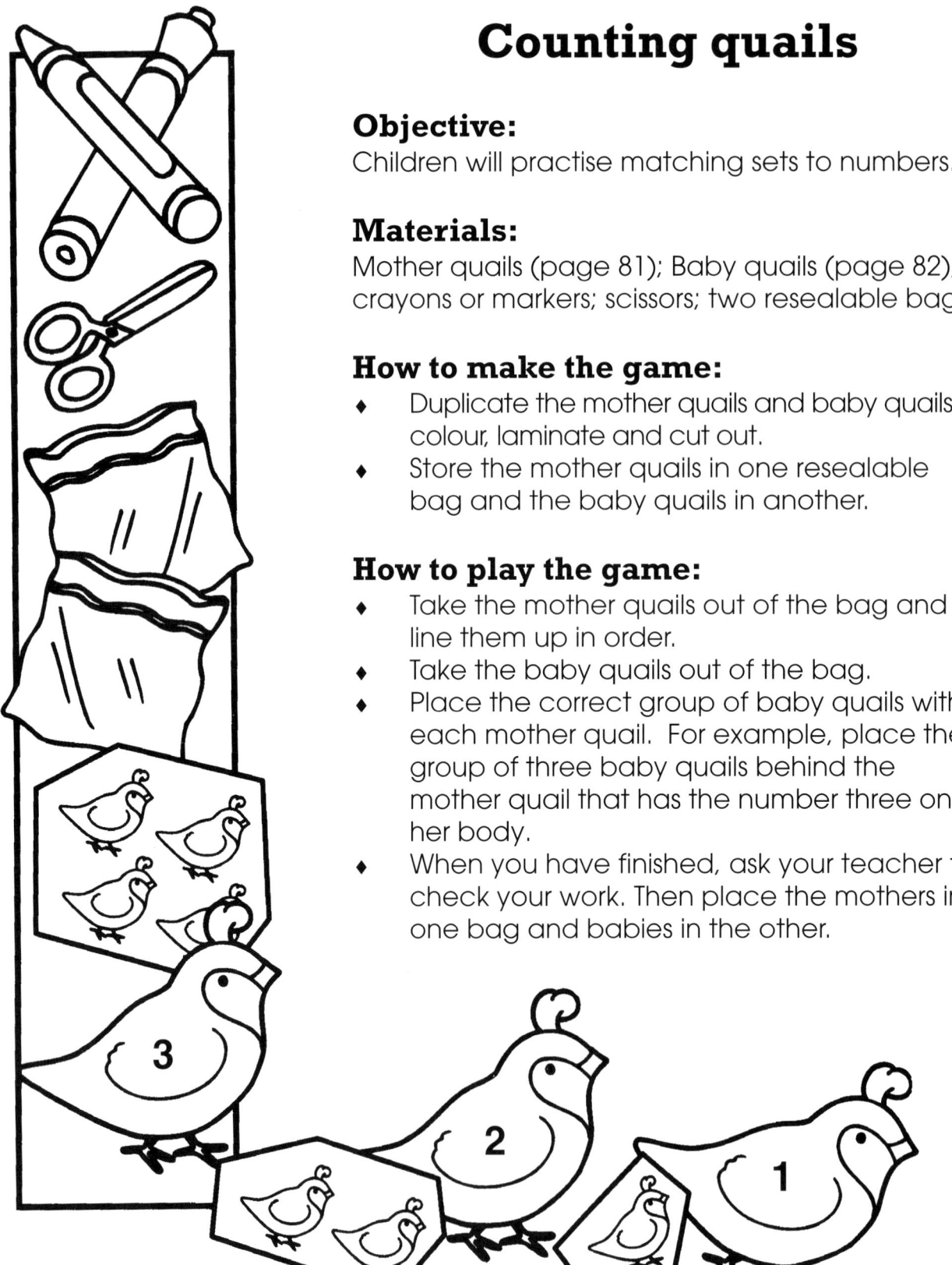

Objective:
Children will practise matching sets to numbers.

Materials:
Mother quails (page 81); Baby quails (page 82); crayons or markers; scissors; two resealable bags.

How to make the game:
♦ Duplicate the mother quails and baby quails, colour, laminate and cut out.
♦ Store the mother quails in one resealable bag and the baby quails in another.

How to play the game:
♦ Take the mother quails out of the bag and line them up in order.
♦ Take the baby quails out of the bag.
♦ Place the correct group of baby quails with each mother quail. For example, place the group of three baby quails behind the mother quail that has the number three on her body.
♦ When you have finished, ask your teacher to check your work. Then place the mothers in one bag and babies in the other.

Book link:
The Bird Alphabet byJerry Pallotta, Edgar Stewart (Charles Bridge Publishing, 1999). Introduces the letters A to Z by describing birds from around the world.

Mother quails

Baby quails

Brilliant Publications – www.brilliantpublications.co.uk
A – Z Maths Games by Karen M. Breitbart

Rabbits and carrots

Objective:
Children will find solutions to simple addition problems.

Materials:
Rabbits (page 84); Carrots (page 85); crayons or markers; scissors; counters; 2 resealable bags.

How to make the game:
♦ Duplicate the rabbits and carrots, colour, laminate and cut out.
♦ Store the rabbits and carrots in one resealable bag and the counters in another.

How to play the game:
♦ Take the rabbits and carrots out of the bag and line them up in two rows.
♦ Solve the addition problem on each rabbit. (If you need help, use the counters to find the solution for each problem.)
♦ When you have finished, ask your teacher to check your work. Then place the rabbits and carrots in one bag and the counters in another.

Book link:
The Tale of Peter Rabbit by Beatrix Potter (Warne, 1998). Peter Rabbit goes on an adventure in Mr McGregor's garden.

Rabbits

Brilliant Publications – www.brilliantpublications.co.uk
A – Z Maths Games by Karen M. Breitbart

Carrots

Railway carriages

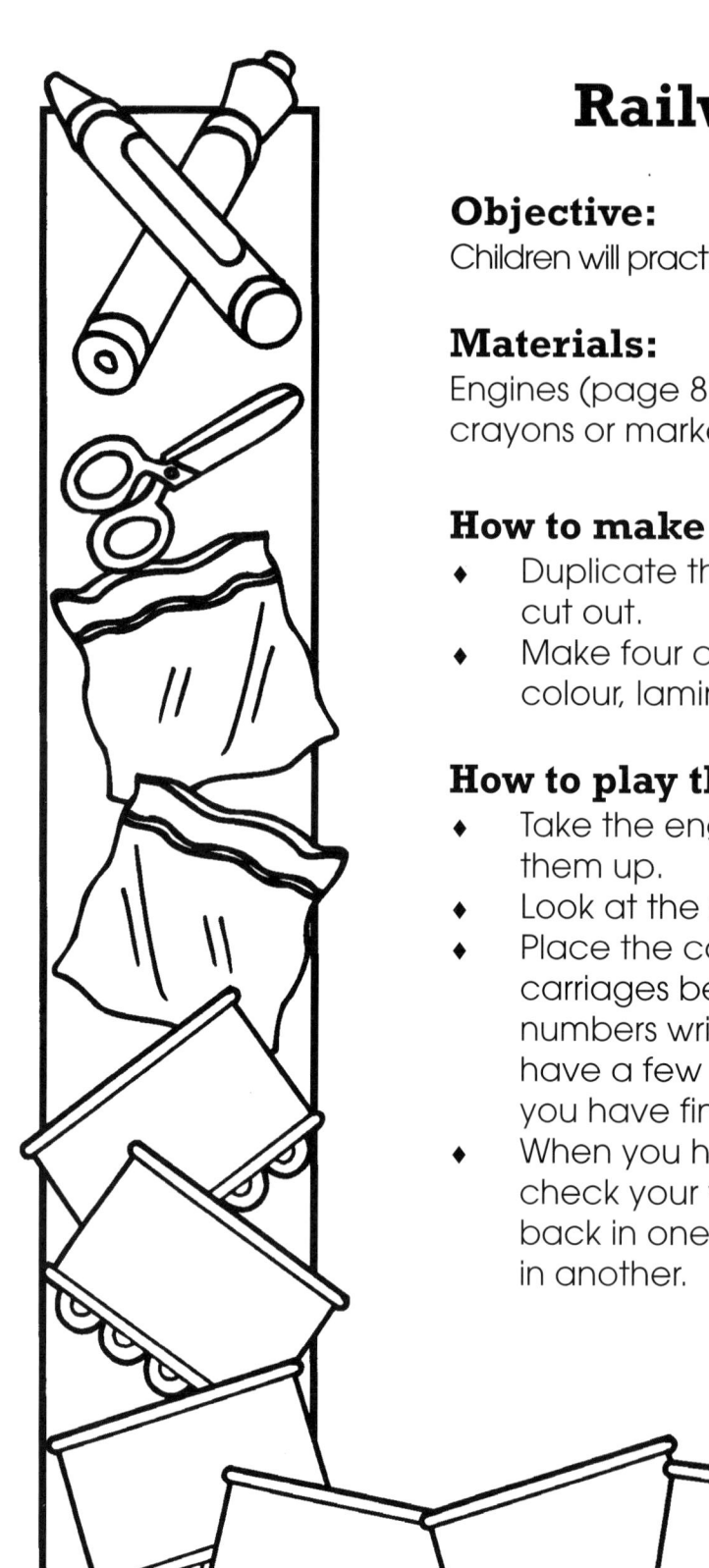

Objective:
Children will practise making sets to match numbers.

Materials:
Engines (page 87); Railway carriages (page 88); crayons or markers; scissors; 2 resealable bags.

How to make the game:
♦ Duplicate the engines, colour, laminate, and cut out.
♦ Make four copies of the railway carriages, colour, laminate and cut out.

How to play the game:
♦ Take the engines out of the bag and line them up.
♦ Look at the numbers on each engine.
♦ Place the correct number of railway carriages behind each engine to match the numbers written on the engines. (You will have a few extra railway carriages when you have finished.)
♦ When you have finished, ask your teacher to check your work. Then place the engines back in one bag and the railway carriages in another.

Book link:
The Caboose Who Got Loose by Bill Peet (Houghton Mifflin, 1980). Katy Caboose wishes she were almost anything but a caboose. When Katy's bolt breaks, she is free.

Engines

Railway carriages

Brilliant Publications – www.brilliantpublications.co.uk
A – Z Maths Games by Karen M. Breitbart

Spaceships and stars

Objective:
Children will find solutions to simple subtraction problems.

Materials:
Spaceships (page 90); Stars (page 91); crayons or markers; scissors; counters; 2 resealable bags.

How to make the game:
- Duplicate the spaceships and stars, colour, laminate and cut out.
- Store the spaceships and stars in one bag and the counters in the other.

How to play the game:
- Take the spaceships out of the bag and line them face up on a table.
- Solve the subtraction problem on each spaceship. (Use the counters to help solve the problem.)
- Match each spaceship problem to the star that has the correct answer.
- When you have finished, ask your teacher to check your work. Then place the space ships and stars back in one bag and the counters in the other.

Book link:
Nora's Stars by Satomi Ichikawa (PaperStar Book, 1997). The stars become Nora's toys until the sky cries, and Nora returns them.

Spaceships

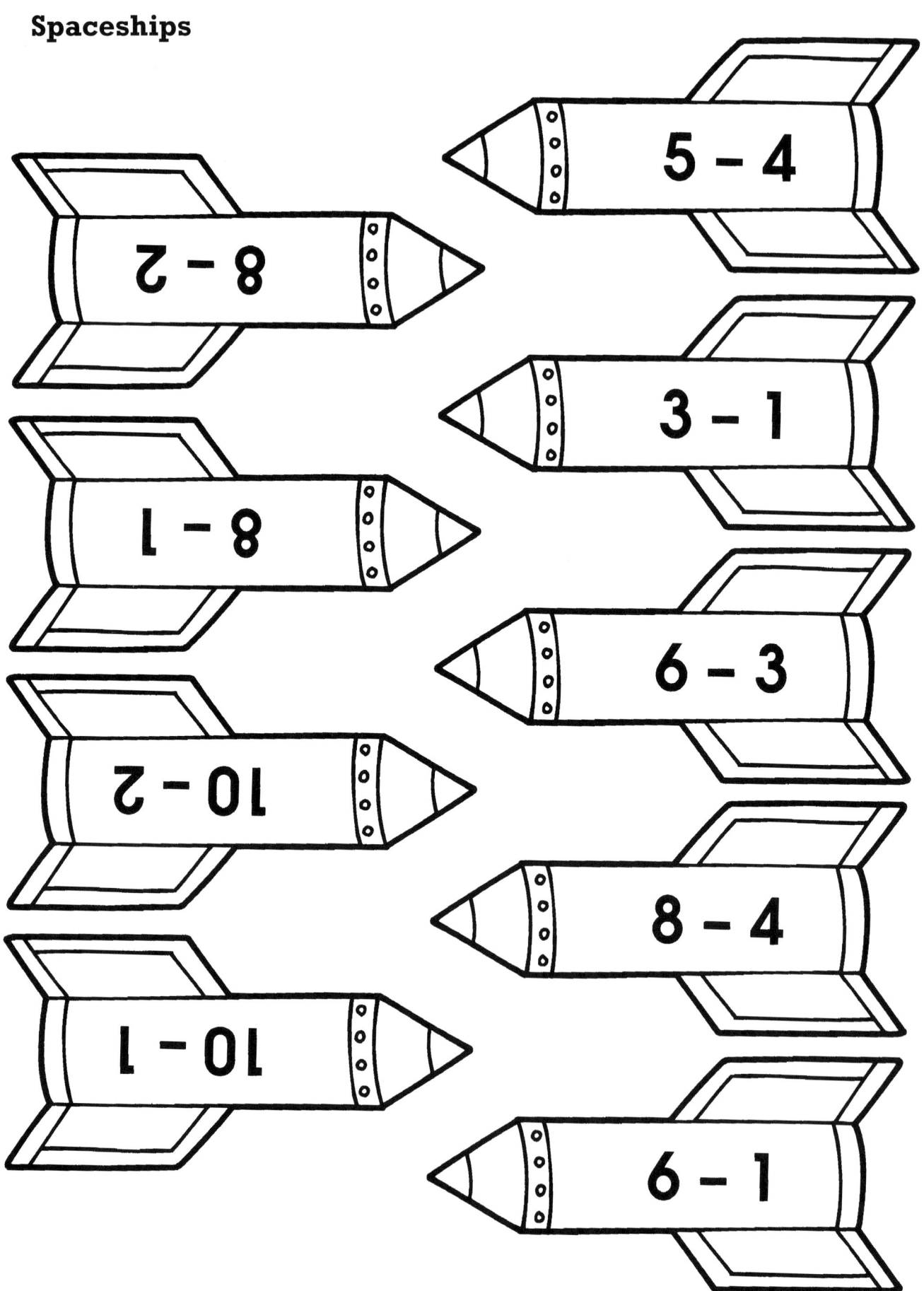

5 − 4

8 − 2

3 − 1

8 − 1

6 − 3

10 − 2

8 − 4

10 − 1

6 − 1

Brilliant Publications – www.brilliantpublications.co.uk
A – Z Maths Games by Karen M. Breitbart

Stars

Subtracting subs

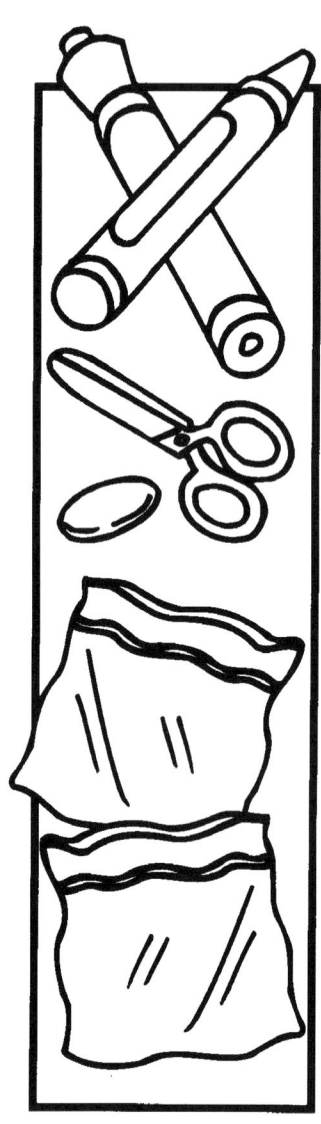

Objective:
Children will find solutions to simple subtraction problems.

Materials:
Submarines (page 93); crayons or markers; scissors; counters; hole punch; 9 small split pins; two resealable bags.

How to make the game:
◆ Duplicate the submarines and periscopes, colour, laminate and cut out.
◆ Punch a hole in each submarine and periscope and attach the correct periscope to the back of each submarine using split pins.
◆ Store the submarines in one resealable bag and the counters in the other.

How to play the game:
◆ Spread the submarines face up on a table.
◆ Look at the subtraction problem on each submarine.
◆ Solve the problems using counters to help you.
◆ When you have finished, check your answers by sliding the periscope up on each submarine. Then slide the periscopes back down.
◆ Place the submarines in one bag and the counters in the other.

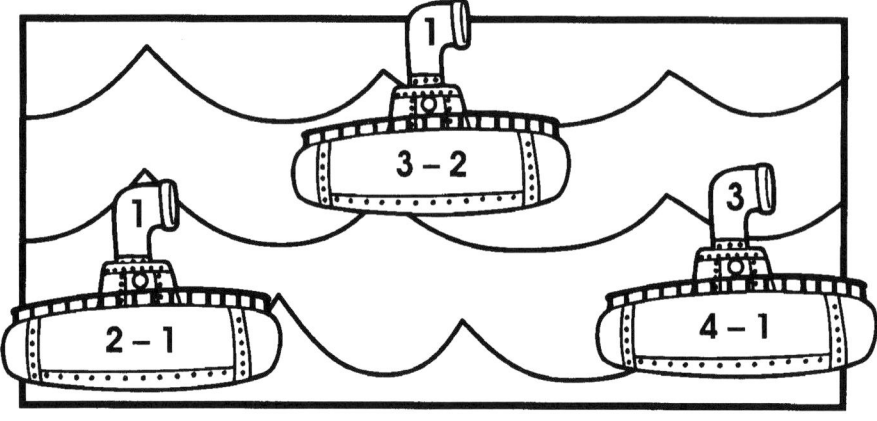

Book link:
Nine naughty Kittens by Linda Jennings, Caroline Jayne Church (Little Tiger Press, 1999). This rhyming, counting book is full of hidden surprises.

Submarines

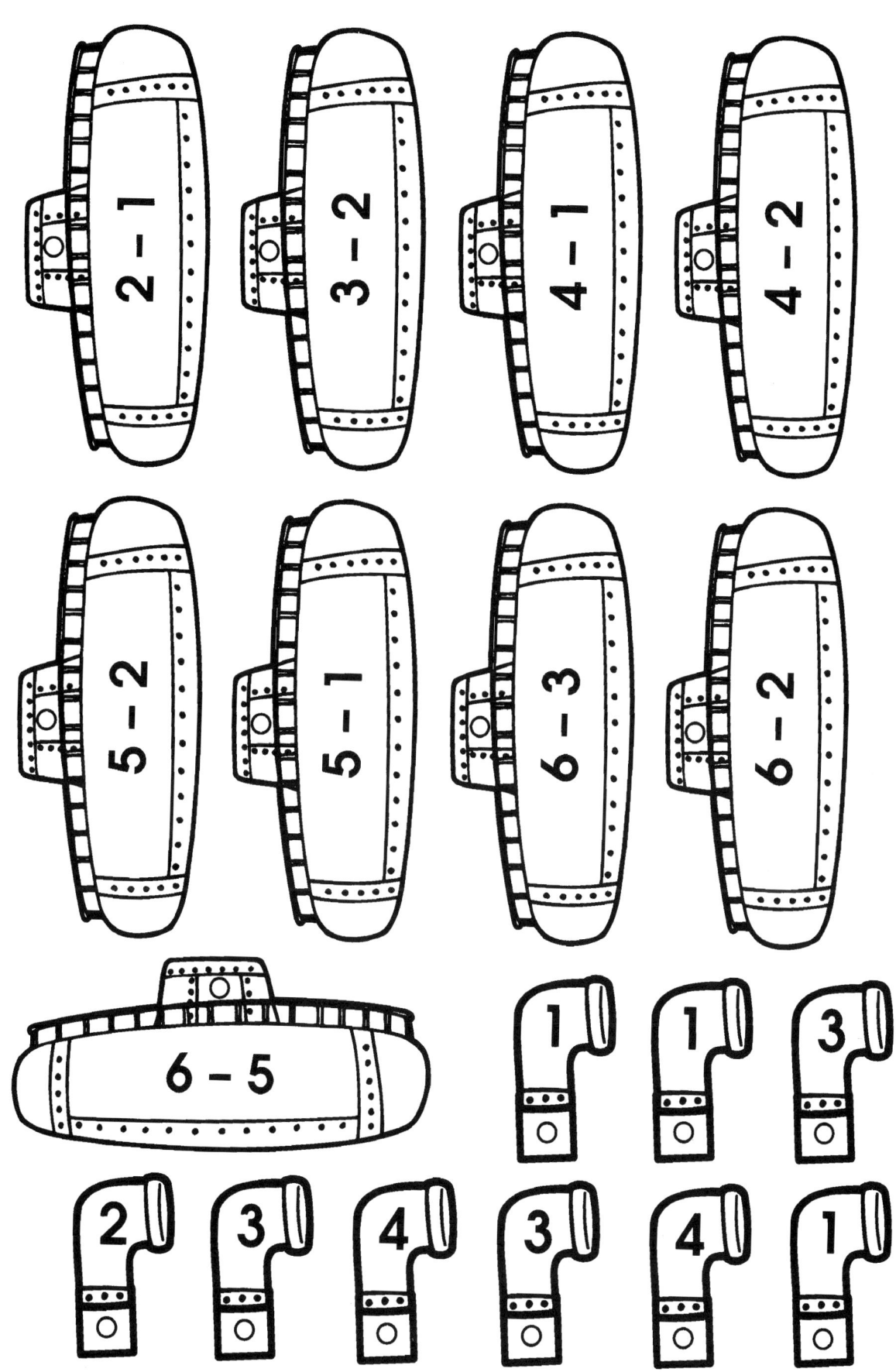

Tiny teeth

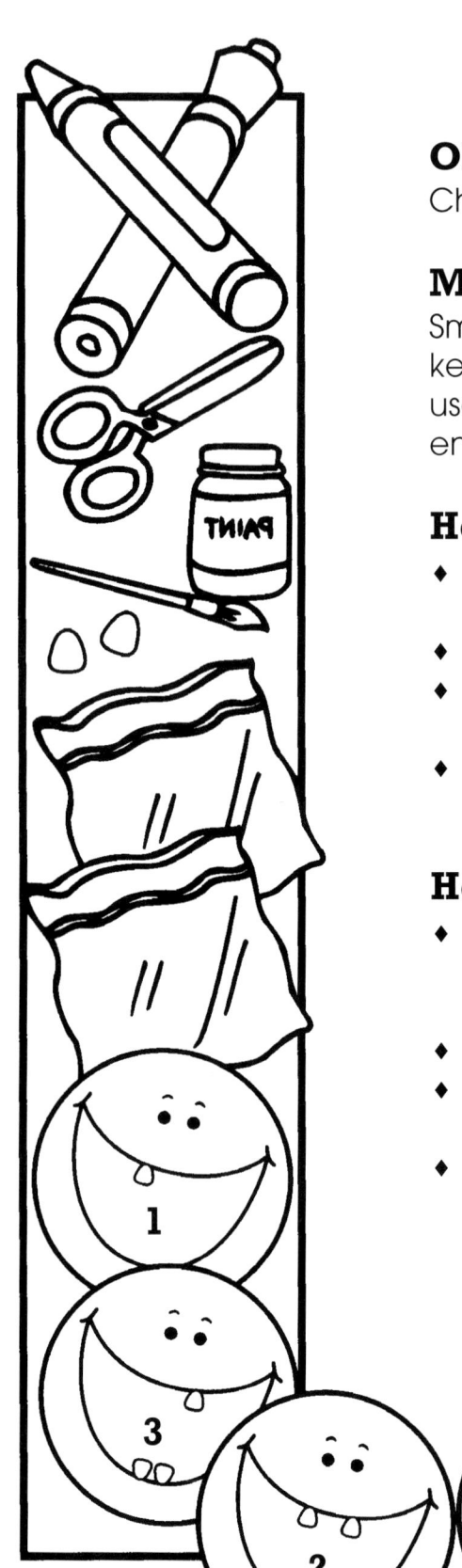

Objective:
Children will make sets of the numbers 1 to 15.

Materials:
Smiley face (page 95); unpopped popcorn kernels (120 or more; white spray paint (for adult use only); crayons or markers; scissors; large envelope; resealable bag.

How to make the game:
♦ Duplicate the smiley face 15 times, colour, laminate and cut out.
♦ Number the smiley faces from 1 to 15.
♦ Paint the unpopped popcorn kernels white and let dry.
♦ Store the smiley faces in the large envelope and the white popcorn kernels in the bag

How to use the game:
♦ Take the smiley faces out of the envelope and line them up in numerical order from 1 to 15.
♦ Look at the number on each smiley face.
♦ Place the correct number of white 'teeth' in the mouth of each smiley face.
♦ When you have finished, ask your teacher to check your work. Then place the smiley faces back in the envelope and the 'teeth' back in the bag.

Book link:
The Tooth Fairy by Peter Collington (Red Fox, 1998). A tooth fairy goes to great lengths when a girl loses a tooth.

Brilliant Publications – www.brilliantpublications.co.uk
A – Z Maths Games by Karen M. Breitbart

Smiley face

Turtle tallying

Objective:
Children will make sets to match numerals from 10 to 20.

Materials:
Turtles (page 97); crayons or markers; scissors; small counters (at least 165); 2 resealable bags.

How to make the game:
- Duplicate the turtles, colour, laminate and cut out. (Enlarge the turtles if you are using large counters.)
- Store the turtles in one resealable bag and the counters in the other.

How to play the game:
- Take the turtles out of the bag and spread them face up in numerical order from 10 to 20.
- Use the counters to make sets on each turtle's shell.
- When you have finished, ask your teacher to check your work. Then place the turtles in one bag and the counters in the other.

Book link:
One Was Johnny. A Counting Book by Maurice Sendak (Harper Collins, 1999). This counting book includes a turtle, as well as more than ten other characters.

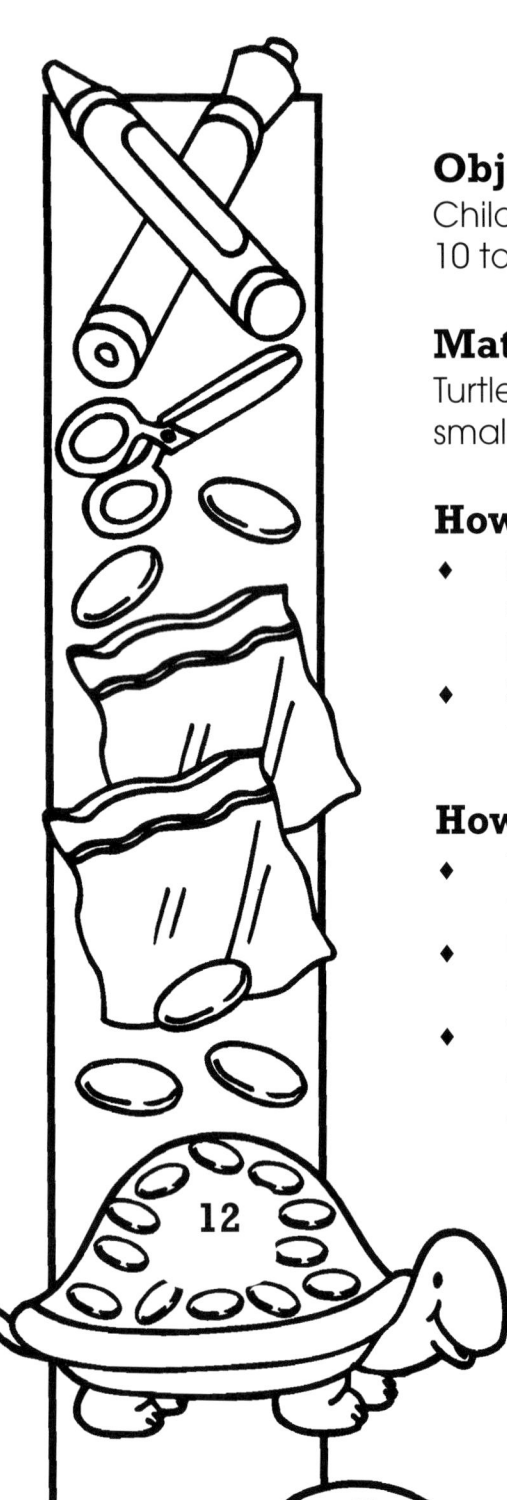

Turtles

Tarantulas' bananas

Objective:
Children will match numbers to sets from 1 to 10.

Materials:
Tarantulas (page 99); Bananas (page 100); crayons or markers; scissors; resealable bag.

How to make the game:
♦ Duplicate the tarantulas and bananas, colour, laminate and cut out.
♦ Store all game pieces in a resealable bag.

How to play the game:
♦ Take the tarantulas and bananas out of the bag and line them face up on a table.
♦ Match the number on each tarantula to the correct group of bananas. For example, if the number on the tarantula is 8, match it with the group of eight bananas.
♦ When you have finished, ask your teacher to check your work. Then place all game pieces back in the bag.

Option:
A tarantula, with legs extended, can be up to 25.5 cm across. Ask the children to measure this distance and make their own life-size tarantulas from paper and pipe cleaners.

Book link:
The Very Busy Spider by Eric Carle (Hamish Hamilton, 1996). Spinning her web keeps the spider busy all day long. A number of animals try to distract the spider from her tasks but the spider ignores them as she is too busy spinning.

Tarantulas

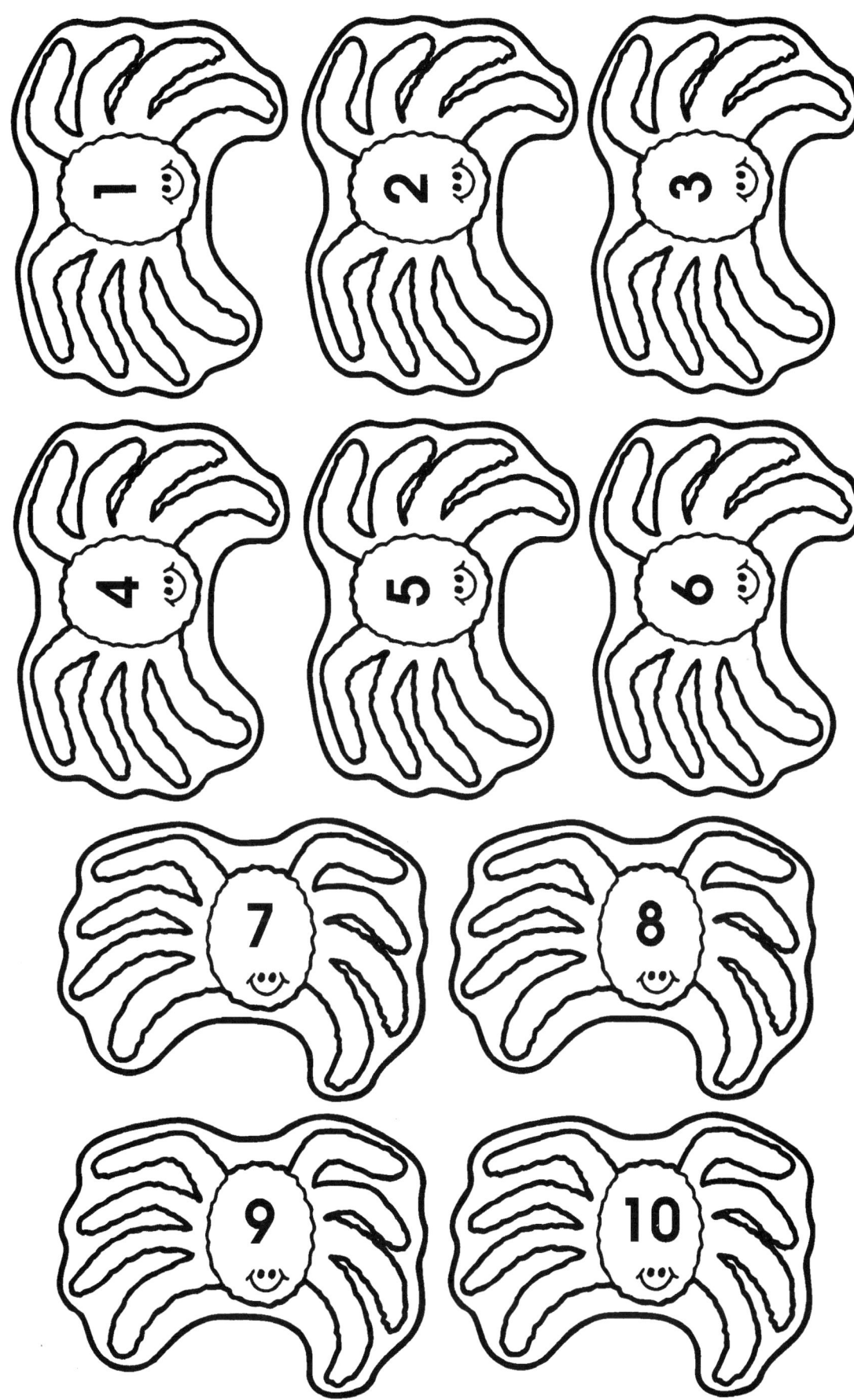

Bananas

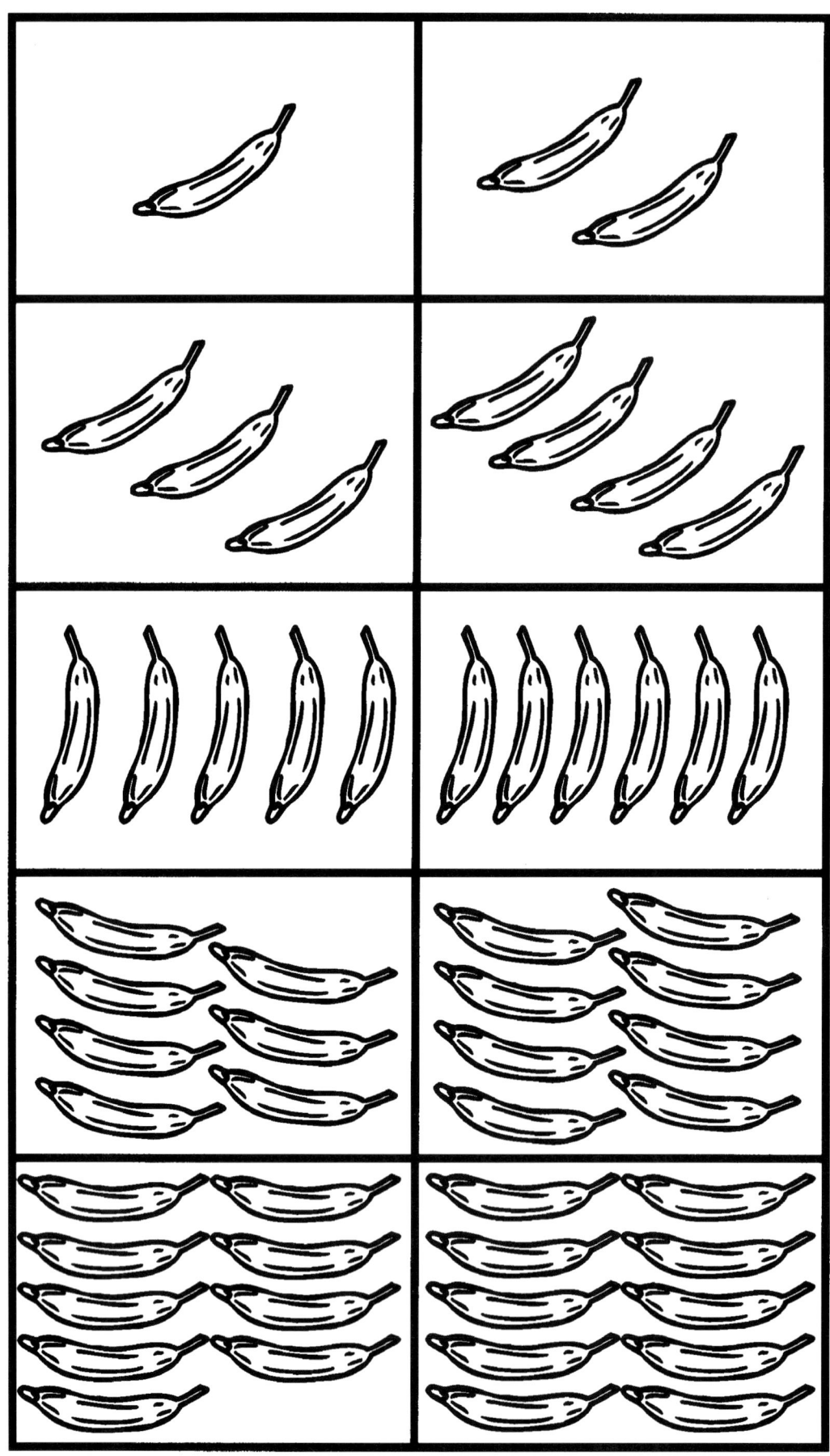

Brilliant Publications – www.brilliantpublications.co.uk
A – Z Maths Games by Karen M. Breitbart

Umbrella game

Objective:
Children will recognise the numbers 1 to 15 and make sets to represent these numbers.

Materials:
Umbrellas (page 102); crayons or markers; scissors; small blue beads (at least 120); 2 resealable bags.

How to make the game:
♦ Duplicate the umbrellas, colour, laminate and cut out.
♦ Store the umbrellas in one resealable bags and the blue beads in the other.

How to play the game:
♦ Spread the umbrellas face up on a table in numerical order from 1 to 15.
♦ Make a matching number set of 'raindrops' for the number on each umbrella. (Place the 'raindrops' above the umbrellas.)
♦ When you have finished, ask your teacher to check your work. Then place the umbrellas back in one bag and the blue 'raindrops' in the other.

Option:
Use cocktail umbrellas and number each one from 1 to 15. These are inexpensive and are often available at arts and crafts stores or cake makers and decorations shops.

Book link:
In the Rain with Baby Duck by Amy Hest, illustrated by Jill Barton (Walker Books, 1998). The rain makes Baby Duck very cross and cranky.

Umbrellas

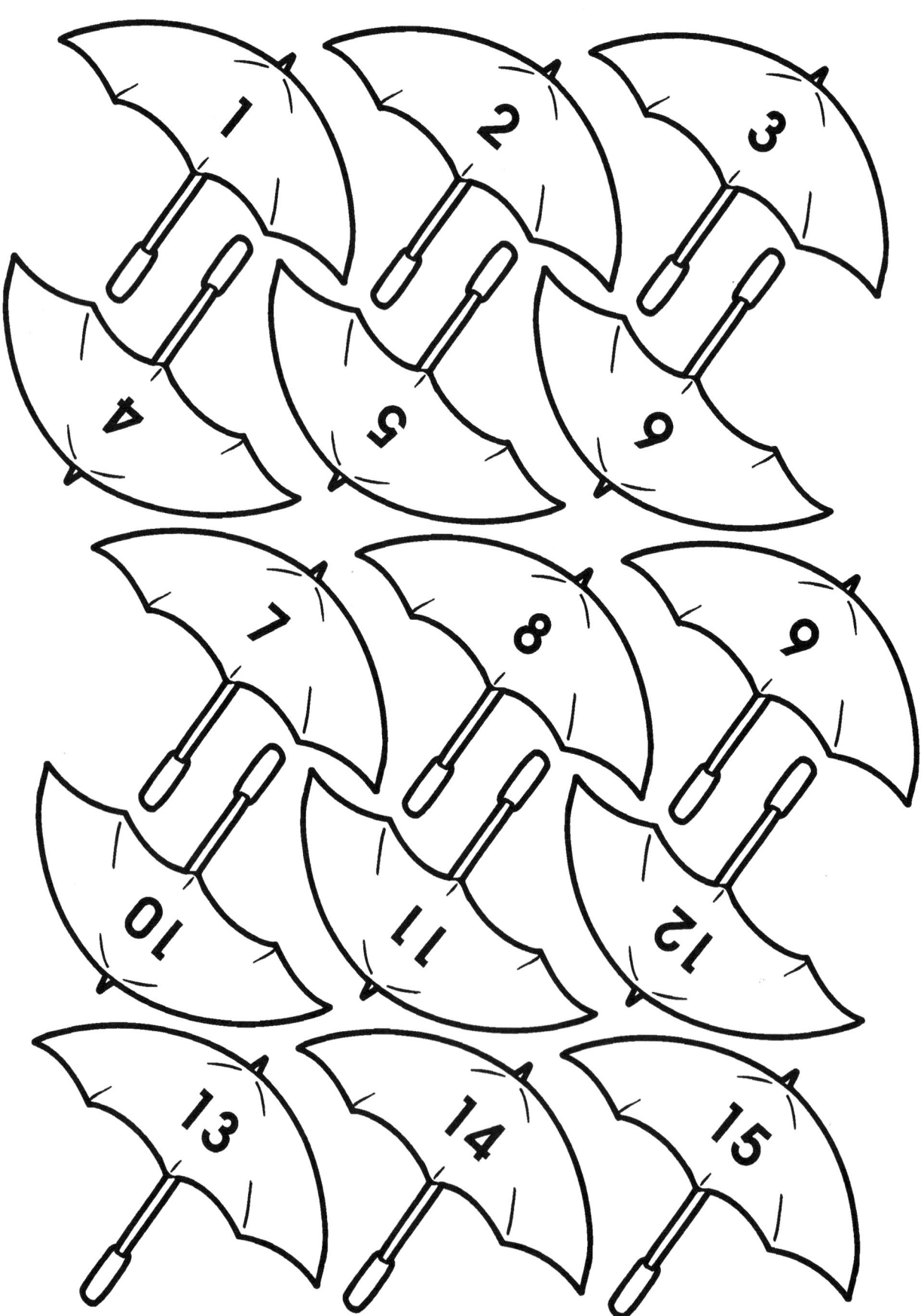

Brilliant Publications – www.brilliantpublications.co.uk
A – Z Maths Games by Karen M. Breitbart

Valentine puzzles

Objective:
Children will match numbers to corresponding sets of objects.

Materials:
Valentines (page 104); crayons or markers; scissors; resealable bag.

How to make the game:
♦ Duplicate the valentines, colour, laminate, cut out and cut in half.
♦ Store the valentines in a resealable bag.

How to play the game:
♦ Take the game pieces out of the bag.
♦ Line the numbered valentine halves in a row from 1 to 10.
♦ Look at the other valentine halves. Each one has a set representing a number.
♦ Match the valentine set halves with the correct valentine number halves.
♦ When you have finished, ask your teacher to check your work. Then place all valentine pieces back in the bag.

Book link:
Guess How Much I Love You by Sam McBratney (Walker Books, 1988). Little Brown Hare thinks of different ways to show Big Brown Hare how much he loves him.

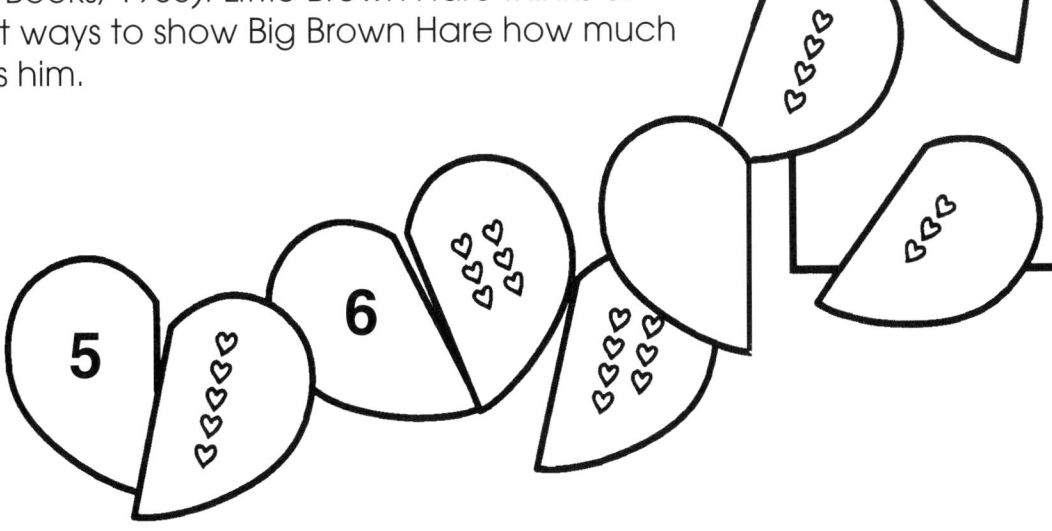

Valentines

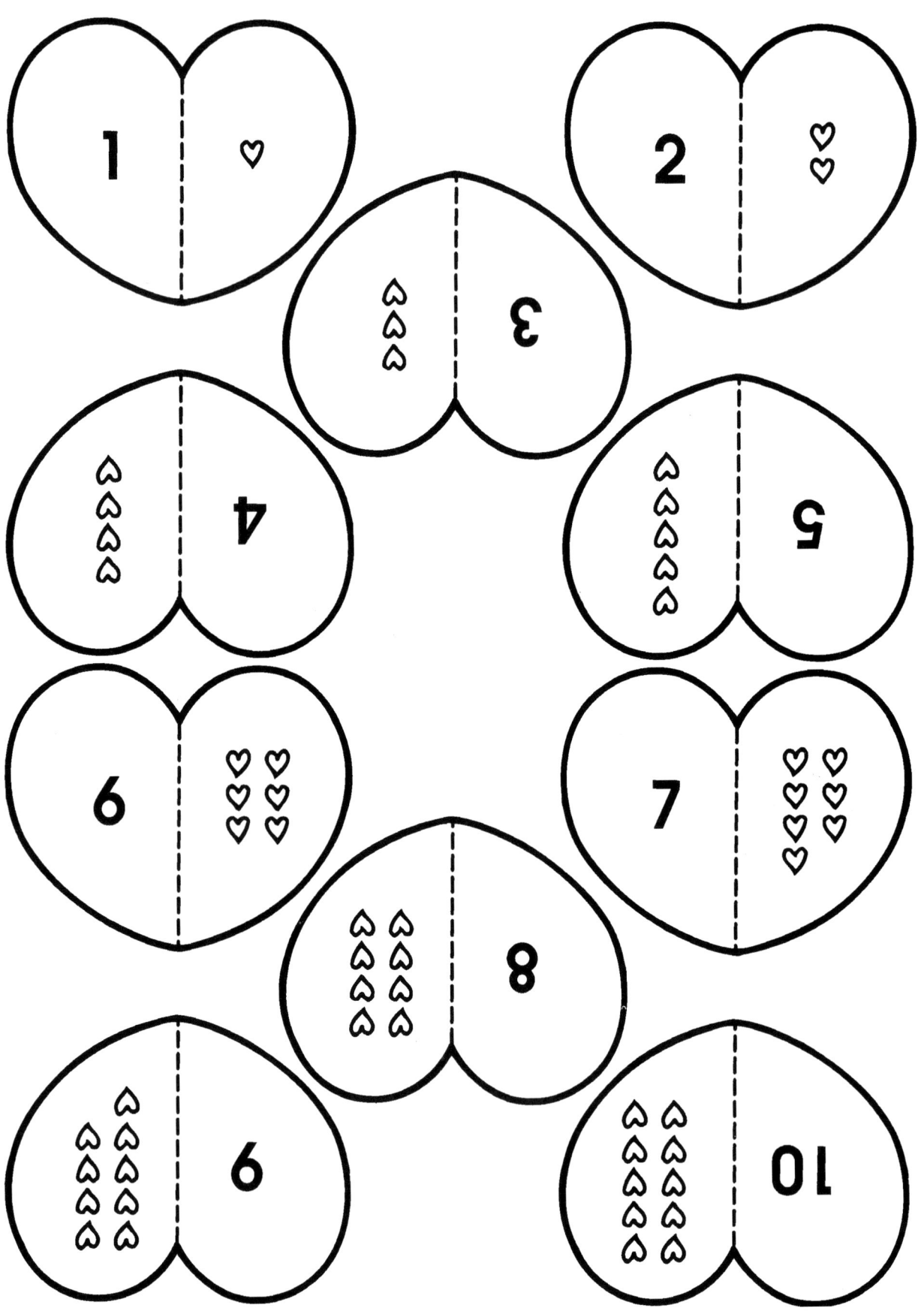

Brilliant Publications – www.brilliantpublications.co.uk
A – Z Maths Games by Karen M. Breitbart

Valentine numbers

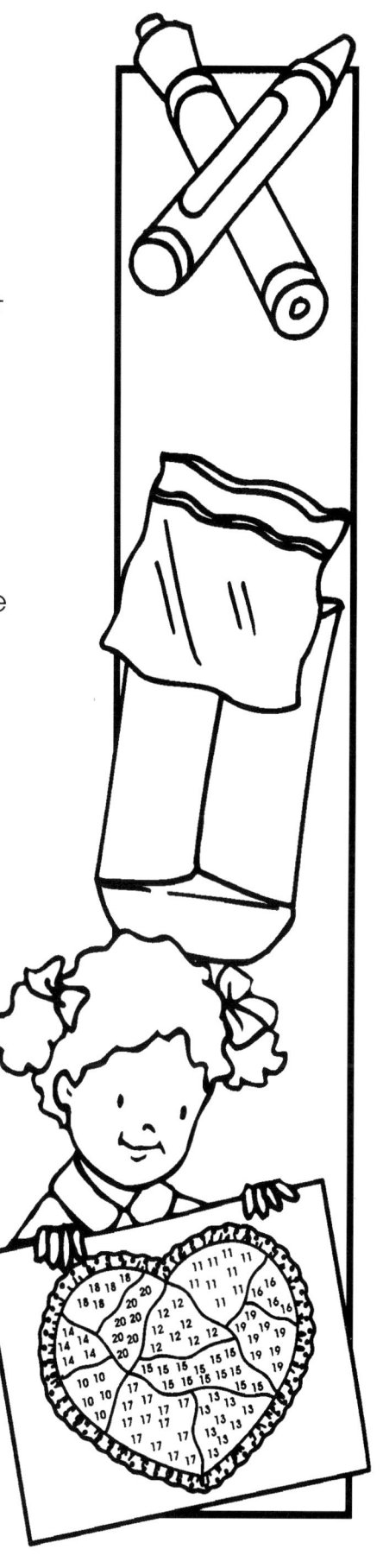

Objective:
Children will practise writing the numbers from 10 to 20.

Materials:
Valentine (page 106); crayons or markers (at least 11 different colours); large envelope; resealable bag.

How to make the game:
♦ Duplicate a copy of the valentine for each child.
♦ Store the valentines in the large envelope.
♦ Store the crayons or markers in the resealable bag.

How to play the game:
♦ Take a valentine out of the large envelope.
♦ Look at the numbers in the different sections of the valentine.
♦ Copy each number over and over to fill each section.
♦ When you have finished, give your picture to the teacher and place the crayons or markers back in the bag.

Option:
Post the completed valentines on a bulletin board in the classroom. Or let the children give these pictures to their friends or relatives on Valentine's Day.

Book link:
Love is a Handful of Honey by Giles Andreae, Vanessa Cabban (Orchard Books, 2000). Mum and Dad listen lovingly as Little Bear re-counts his fun-filled adventures as he settles down to go to sleep.

Valentine

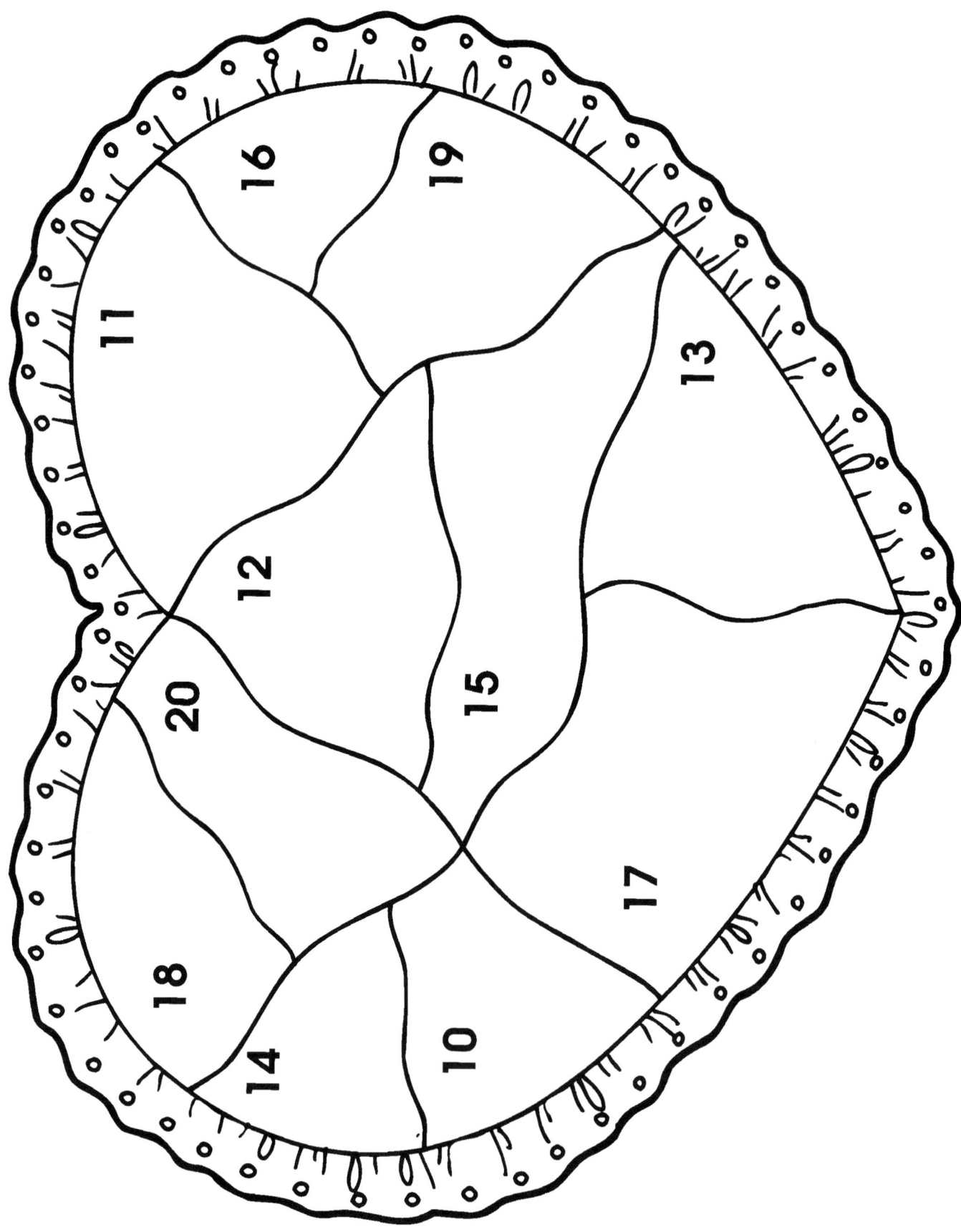

Brilliant Publications – www.brilliantpublications.co.uk
A – Z Maths Games by Karen M. Breitbart

Vampire bats

Objective:
Children will practise making sets of numbers from 1 to 10.

Materials:
Vampire bats (page 108); 10 envelopes; crayons or markers; scissors; resealable bag.

How to make the game:
♦ Number the envelopes from 1 to 10.
♦ Make four copies of the vampire bats, colour, laminate and cut out.
♦ Store the vampire bats and envelopes in a resealable bag.

How to play the game:
♦ Take the envelopes out of the bag and line them up in order. (These envelopes represent caves.)
♦ Look at the number on the outside of each cave.
♦ In each cave, place the number of bats to equal the number on the outside.
♦ When you have finished, ask your teacher to check your work. Then place all game pieces back in the bag.

Option:
Duplicate one copy of the vampire bats for each child. Let the children glue the bats to sheets of black sugar paper. Ask the children number the bats from 1 to 15.

Book link:
The Owl Who Was Afraid of the Dark by Jill Tomlinson (Mammoth, 2000). Find out what happens when the owl meet others who like the dark.

Vampire bats

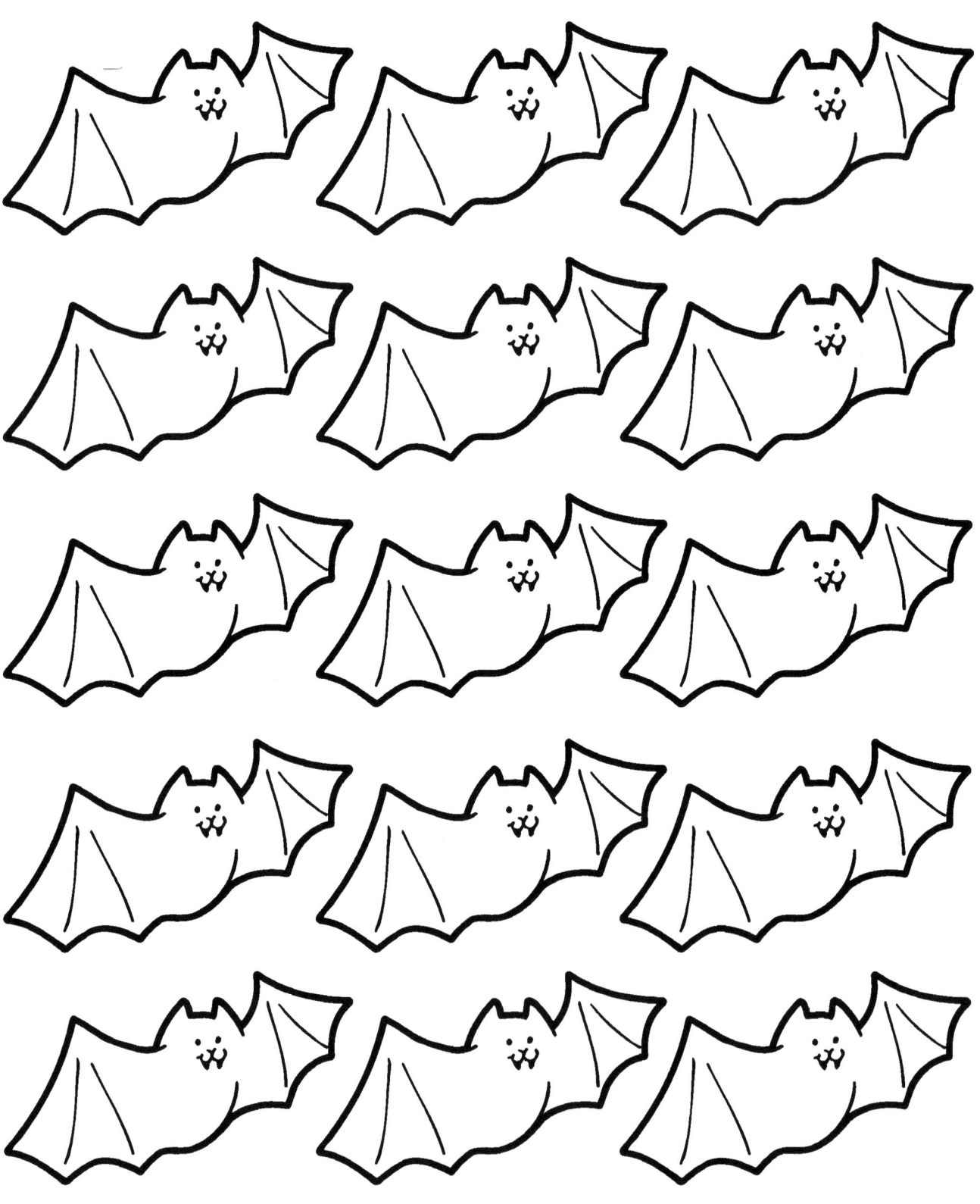

Brilliant Publications – www.brilliantpublications.co.uk
A – Z Maths Games by Karen M. Breitbart

Volcano maths

Objective:
Children will practise making sets to match numbers.

Materials:
Volcanoes (page 110); small red counters; crayons or markers; large envelope; resealable bag.

How to make the game:
♦   Duplicate and enlarge the volcanoes, colour and laminate.
♦   Store the volcanoes in a large envelope.
♦   Store the red counters in the resealable bag.

How to play the game:
♦   Take the sheet of volcanoes out of the large envelope.
♦   Look at the number on each volcano.
♦   Place the correct number of red counters shooting out of each volcano. For example, on the volcano numbered 15, place 15 counters shooting out of the top.
♦   Whe you have finished, ask your teacher to check your work. Then place the volcanoes back in the envelope and the counters in the resealable bag.

Options:
♦   Duplicate a copy of the volcanoes for each child. Ask the children to place stickers or glue sequins on the volcanoes.
♦   Post the pictures on a 'Volcano Maths' bulletin board.

Book Link:
Rock and Rolling – the Earth by Philip Steele (Walker Books, 1997). Discover facts about volcanoes and earthquakes.

Volcanoes

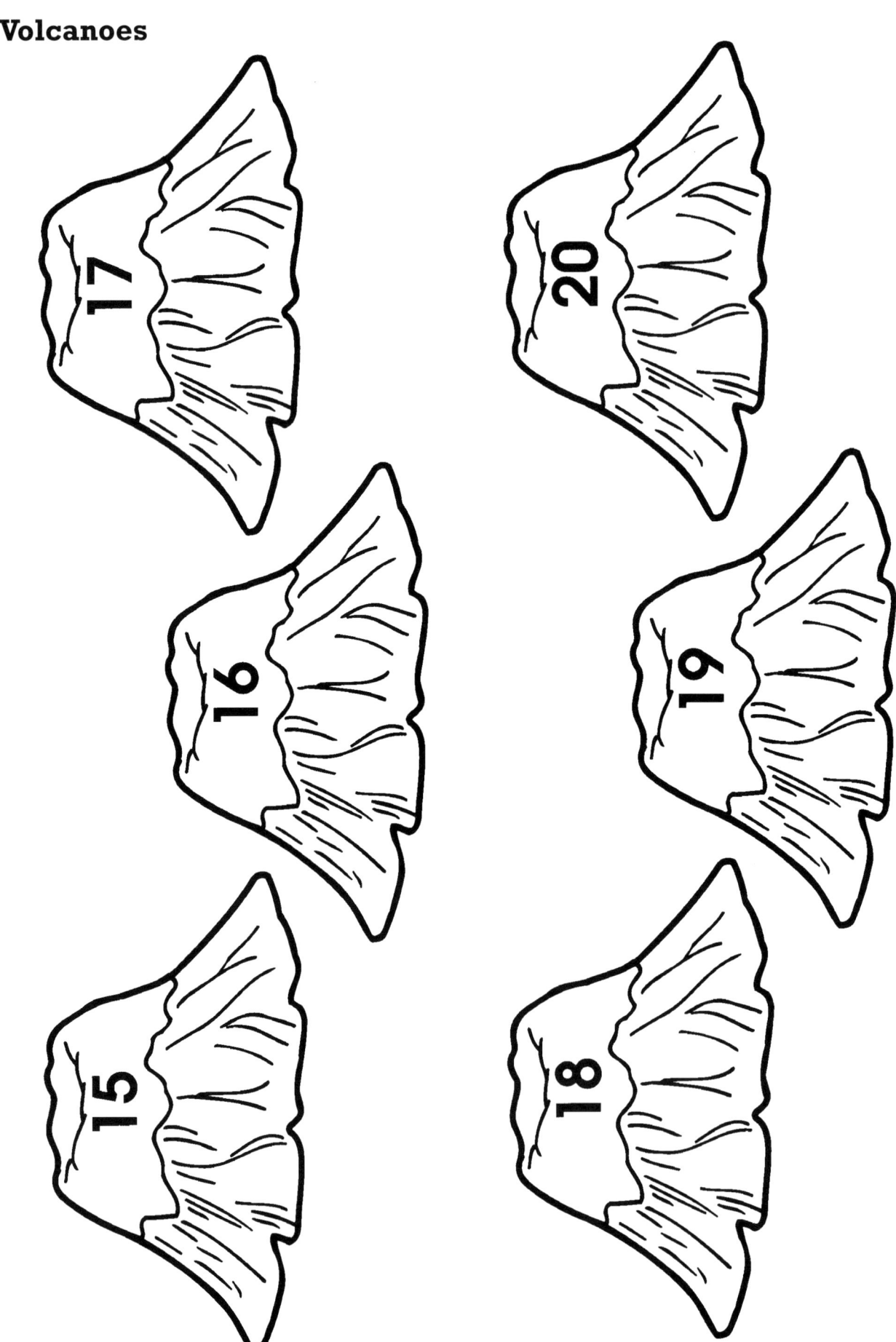

Brilliant Publications – www.brilliantpublications.co.uk
A – Z Maths Games by Karen M. Breitbart

Whales' spouts

Objective:
Children will practise simple subtraction problems.

Materials:
Whales (page 112); Spouts (page 113); crayons or markers; scissors; counters; 2 resealable bags.

How to make the game:
♦ Duplicate the whales and spouts, colour, laminate and cut out.
♦ Store the whales and spouts in one resealable bag and the counters in the other.

How to play the game:
♦ Take the whales out of the bag and line them face up in a row.
♦ Use the counters to solve the subtraction problem on each whale.
♦ For each whale, find the spout that has the correct answer.
♦ When you have matched each whale with a spout, ask your teacher to check your work. Then place all whales and spouts back in one bag and all counters in the other.

Book link:
Jonah and the Whale adapted by Marcia Williams (Walker Books, 1998). A comic-strip retelling of the bible story.

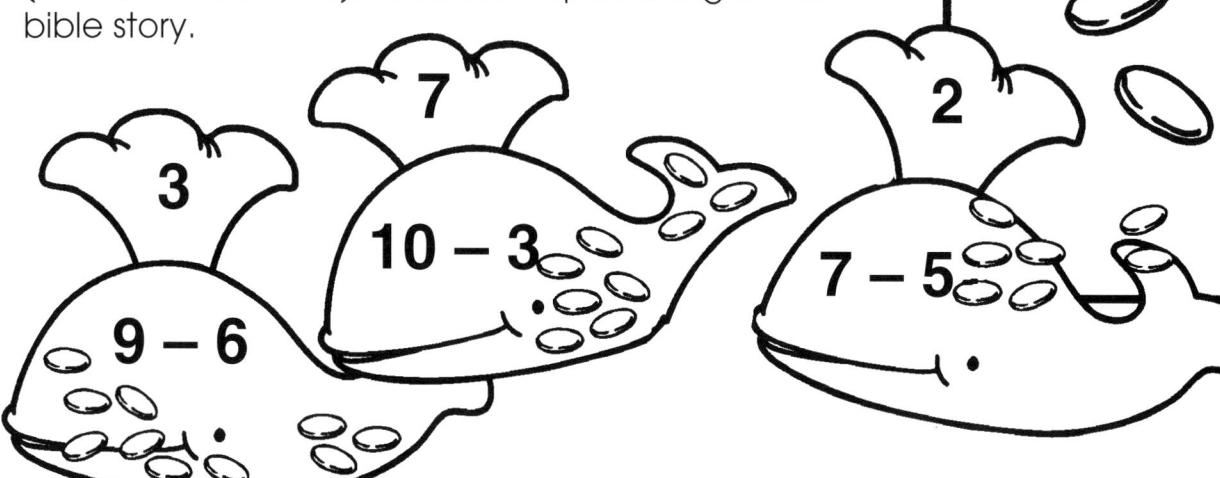

Whales

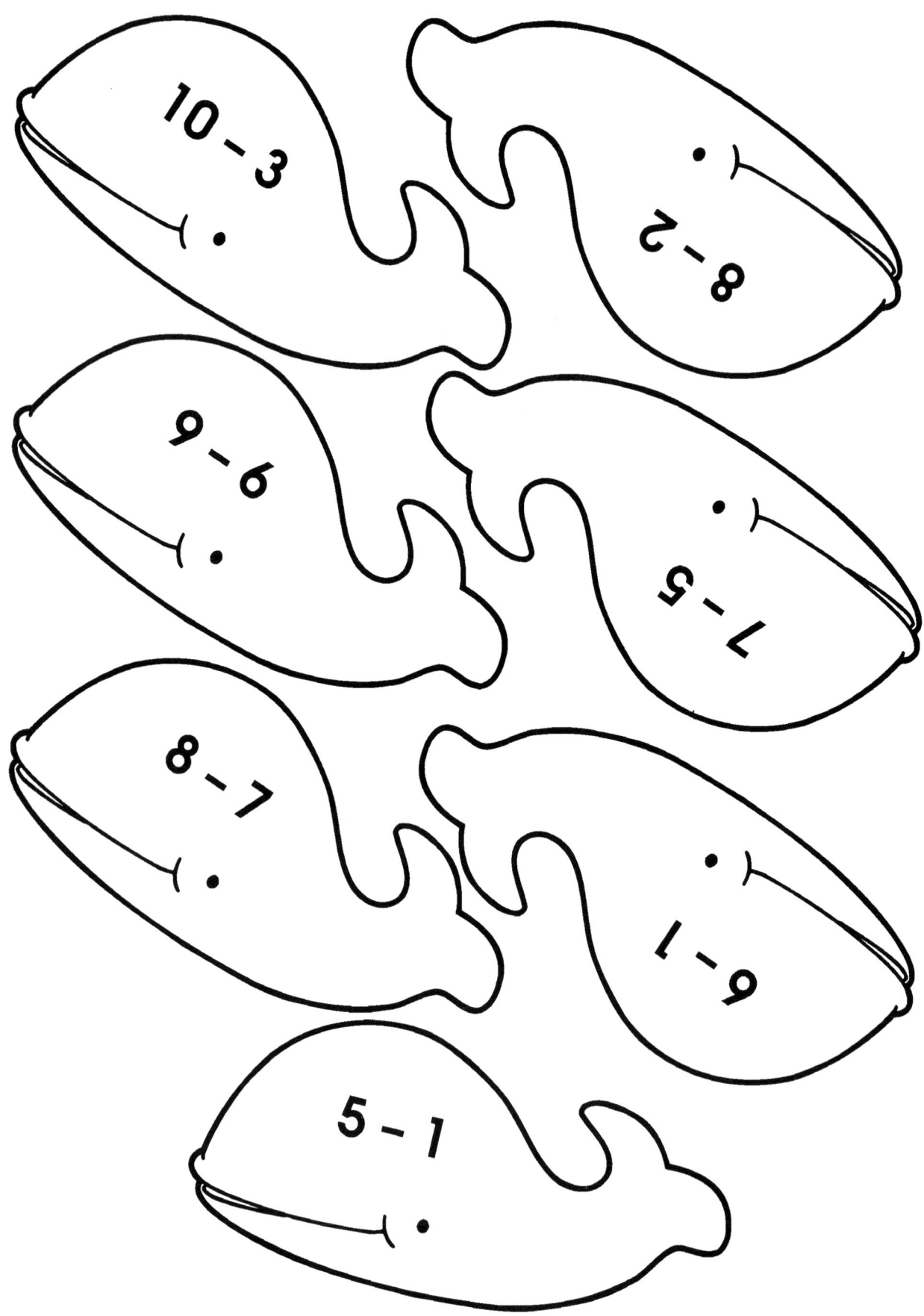

Brilliant Publications – www.brilliantpublications.co.uk
A – Z Maths Games by Karen M. Breitbart

Spouts

Water-lily maths

Objective:
Children will match sets to numbers.

Materials:
Water lilies (page 115); Frogs (page 116; crayons or markers; scissors; large envelope; resealable bag.

How to make the game:
*   Duplicate the water lilies, colour and laminate.
*   Duplicate the frogs, colour, laminate and cut out.
*   Store the water lilies in a large envelope and the frogs in a resealable bag.

How to play the game:
*   Take the water lilies out of the envelope.
*   Look at the number on each water lily.
*   For each water lily, find the frog with the matching number of dots on its back.
*   When you have matched each frog and water lily, ask your teacher to check your work. Then place the water lilies back in the envelope and the frogs back in the bag.

Book Link:
Linnea in Monet's Garden by Christina Bjork, drawings by Lina Anderson (Raben & Sjogren, 1987). This adorable book includes reproductions of Monet's work, as well as photographs of his home and garden at Giverny.

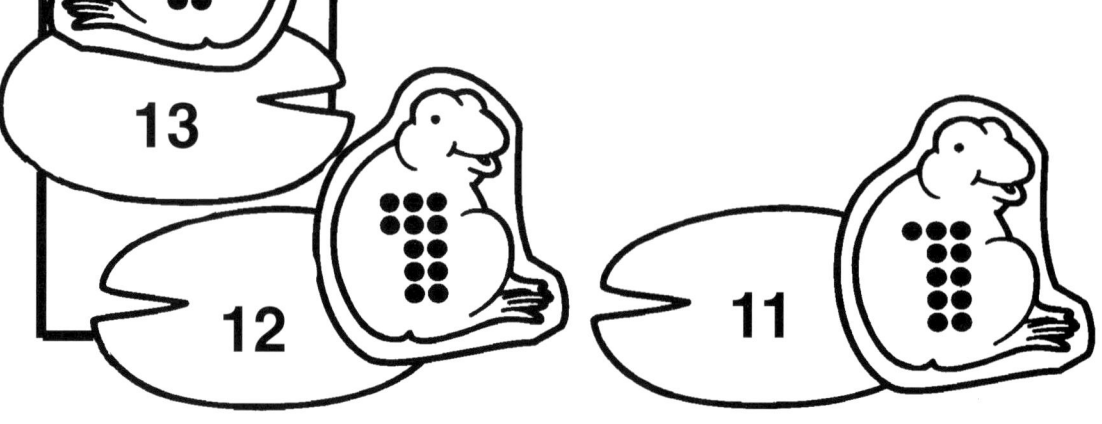

Water lilies

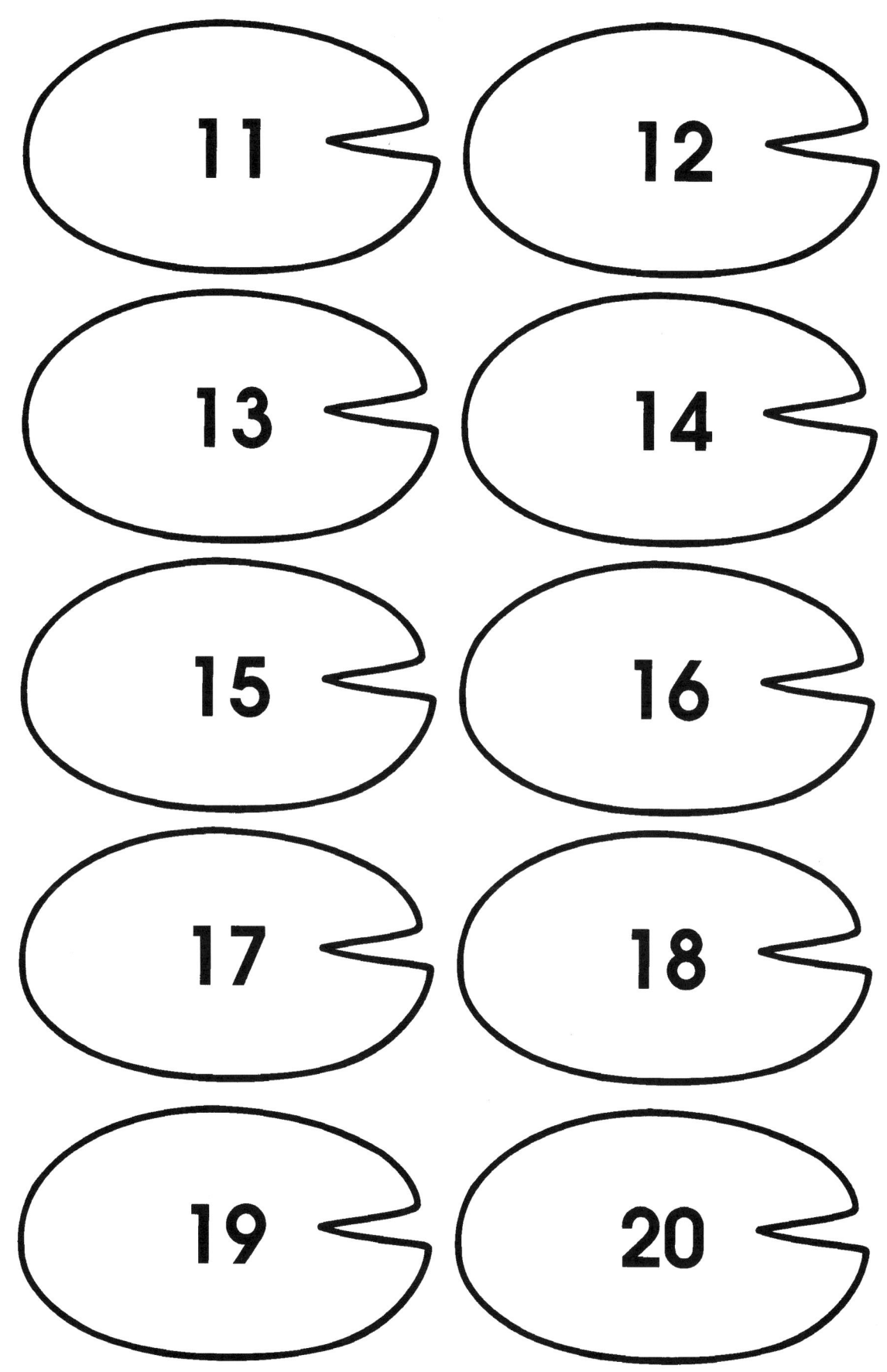

Frogs

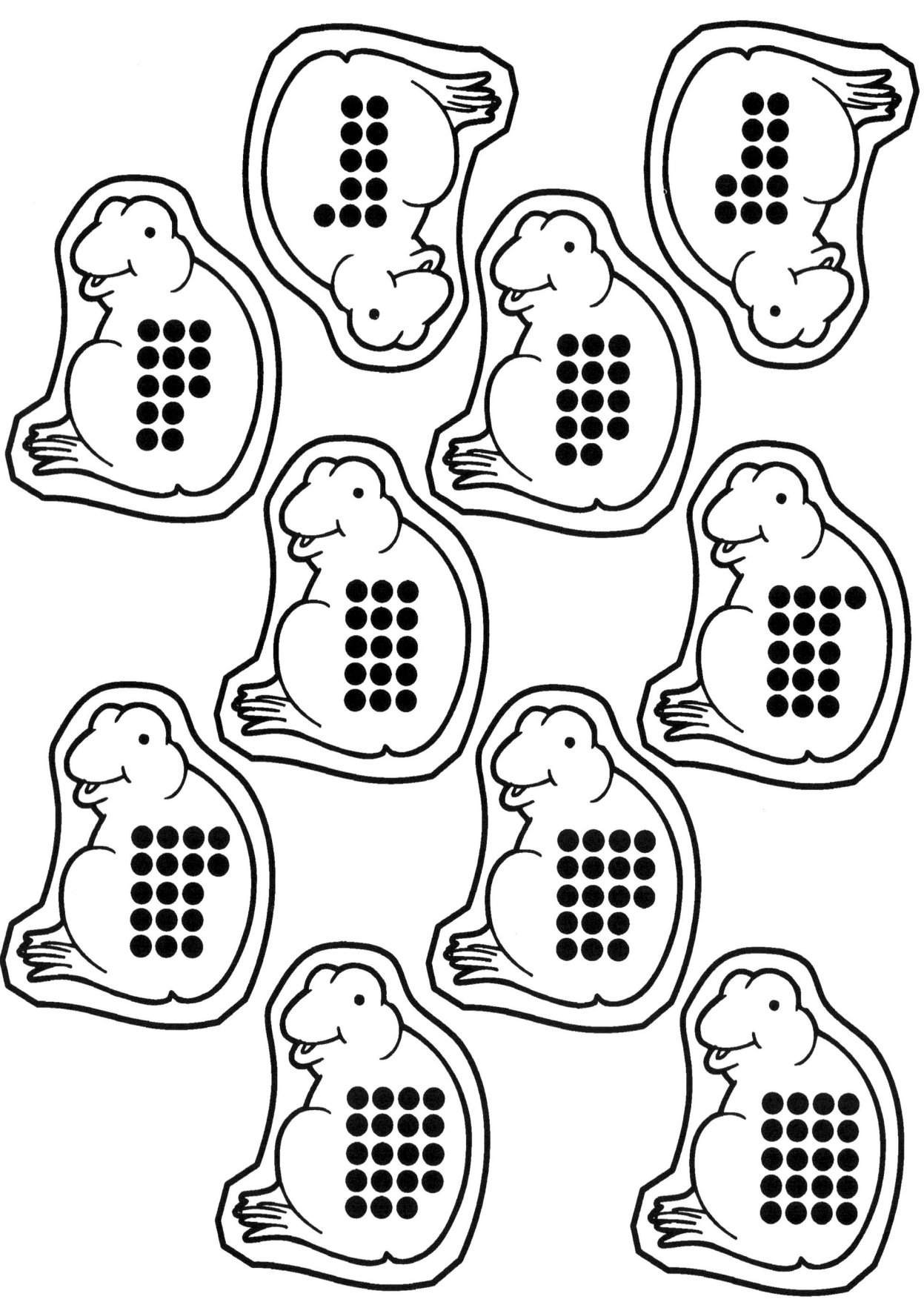

This page may be photocopied by the purchasing institution only.

Brilliant Publications – www.brilliantpublications.co.uk
A – Z Maths Games by Karen M. Breitbart

Wheels

Objective:
Children will practise matching sets to even numbers.

Materials:
Truck Bodies (page 118); Wheels (page 119); crayons or markers; scissors; large envelope.

How to make the game:
♦   Duplicate and enlarge the truck bodies, colour and laminate.
♦   Duplicate two copies of the wheels, laminate and cut out.
♦   Store the truck bodies and the wheels in the envelope.

How to play the game:
♦   Take the truck bodies and wheels out of the envelope.
♦   Look at the number on each truck body.
♦   Place the matching number of wheels on each truck.
♦   When you have finished, ask your teacher to check your work. Then place the game pieces back in the envelope.

Option:
Duplicate a copy of the truck bodies for each child to glue to a sheet of sugar paper. Ask the children to draw wheels on the trucks to match the number on each truck body.

Book link:
Things That Go by Anne Rockwell (Dutton, 1991). This book is divided into sections, including things that go on the road, on the water, in the air and more.

Truck bodies

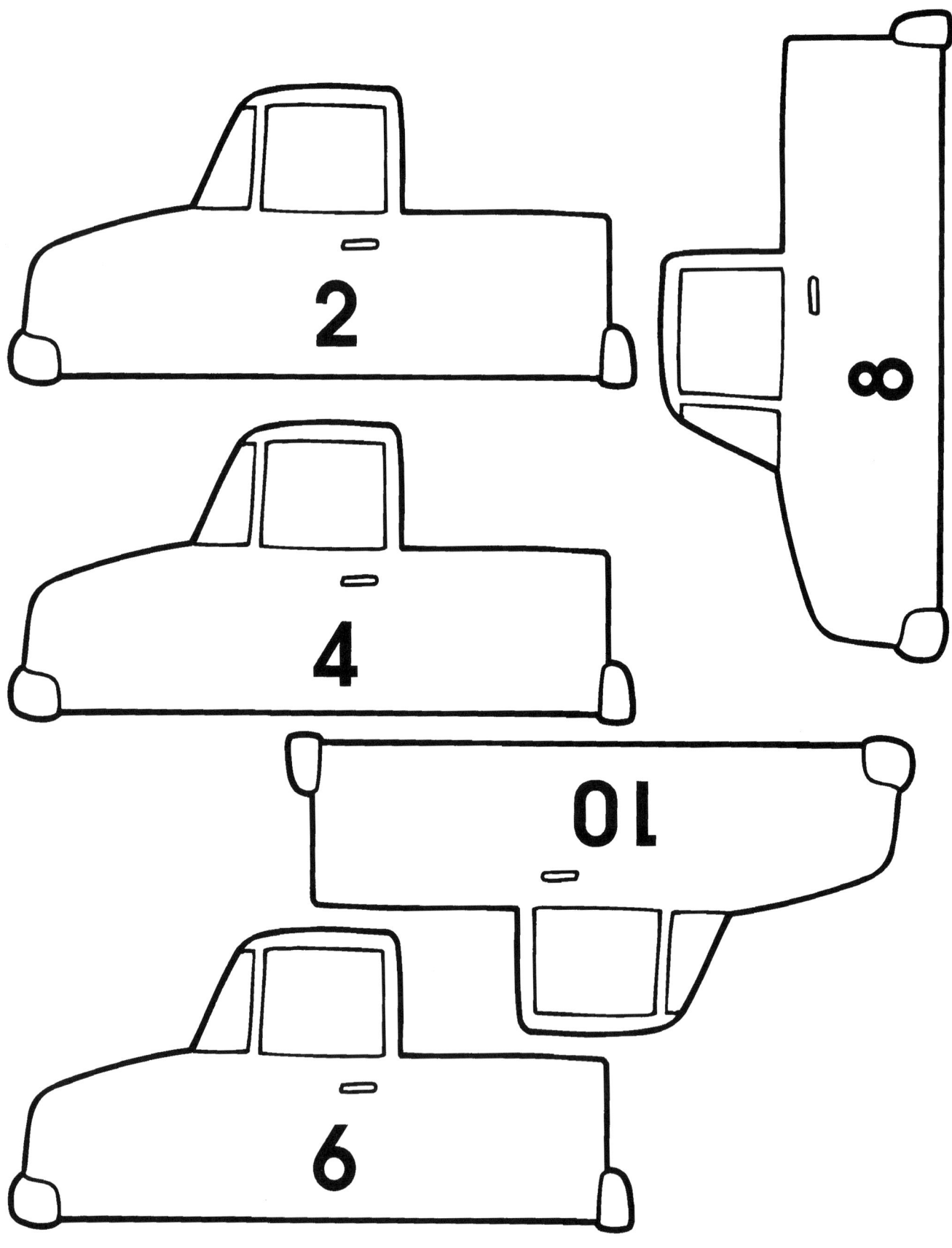

Brilliant Publications – www.brilliantpublications.co.uk
A – Z Maths Games by Karen M. Breitbart

Wheels

Xylophone

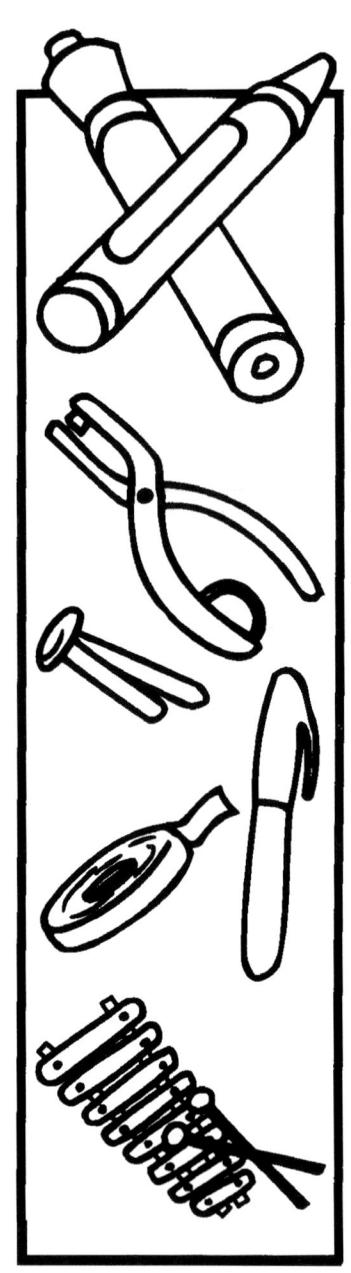

Objective:
Children will practise counting forwards and backwards.

Materials:
Spinner and arrow (page 121); hole punch; split pin; crayons or markers; child's xylophone; masking tape; permanent marker.

How to make the game:
♦ Duplicate the spinner and arrow, colour, laminate and cut out.
♦ Punch a hole in the centre of the spinner and attach the arrow using the split pin.
♦ Use masking tape and a permanent marker to number the xylophone keys in order from 1 to 15.
♦ Store the xylophone and spinner in the Maths Centre.

How to play the game:
♦ Spin the spinner and count from 1 to the number spun, striking the xylophone keys as you count.
♦ Count backwards from the number spun to 1, striking the keys as you count.
♦ Spin the spinner and count forwards and backwards at least five times before putting the game away.

Book link:
Zin! Zin! Zin! A Violin by Lloyd Moss, illustrated by Marjorie Priceman (Simon & Schuster, 1995).
Ten musical instruments play in a musical performance.

Spinner and arrow

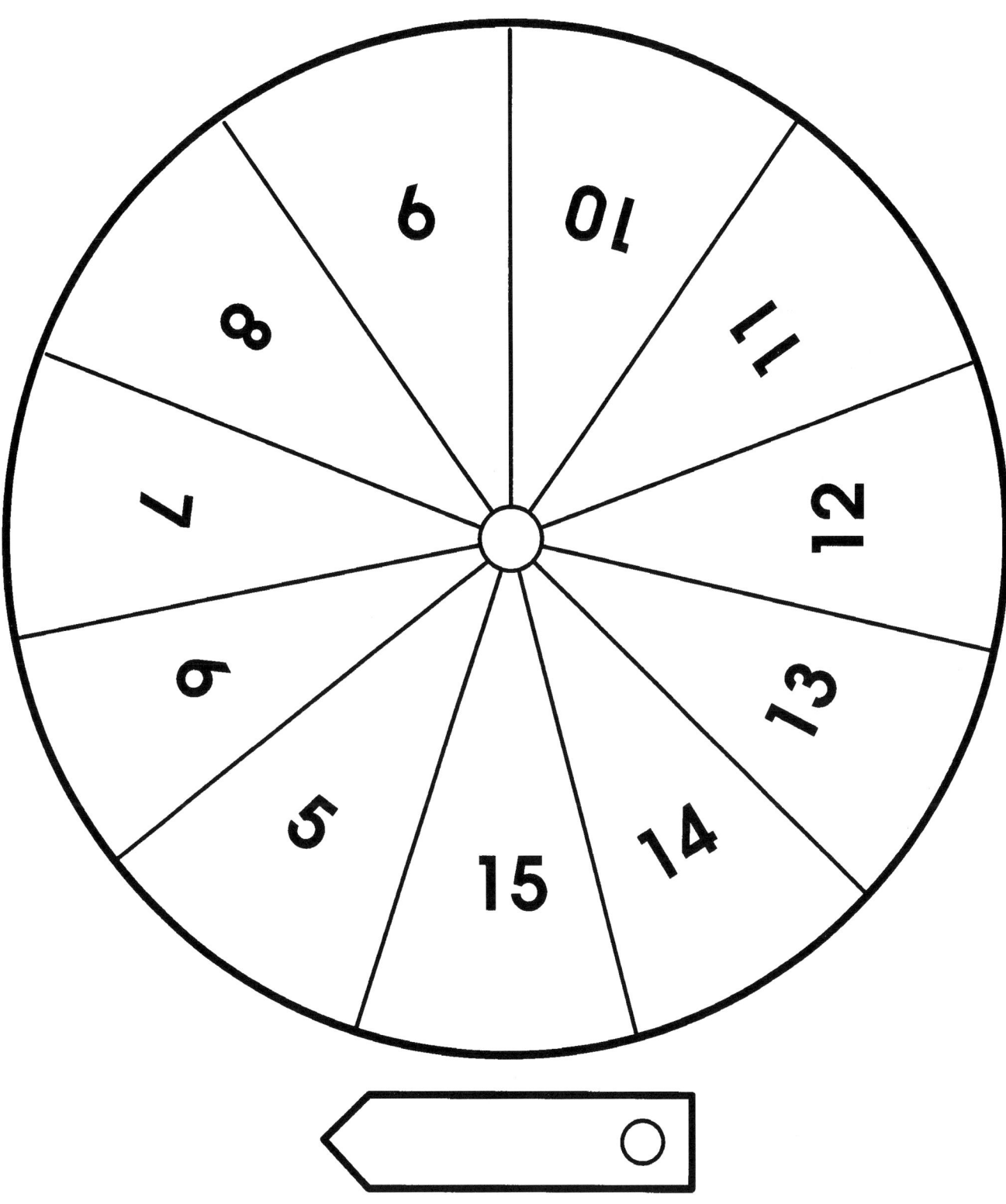

Yolk addition

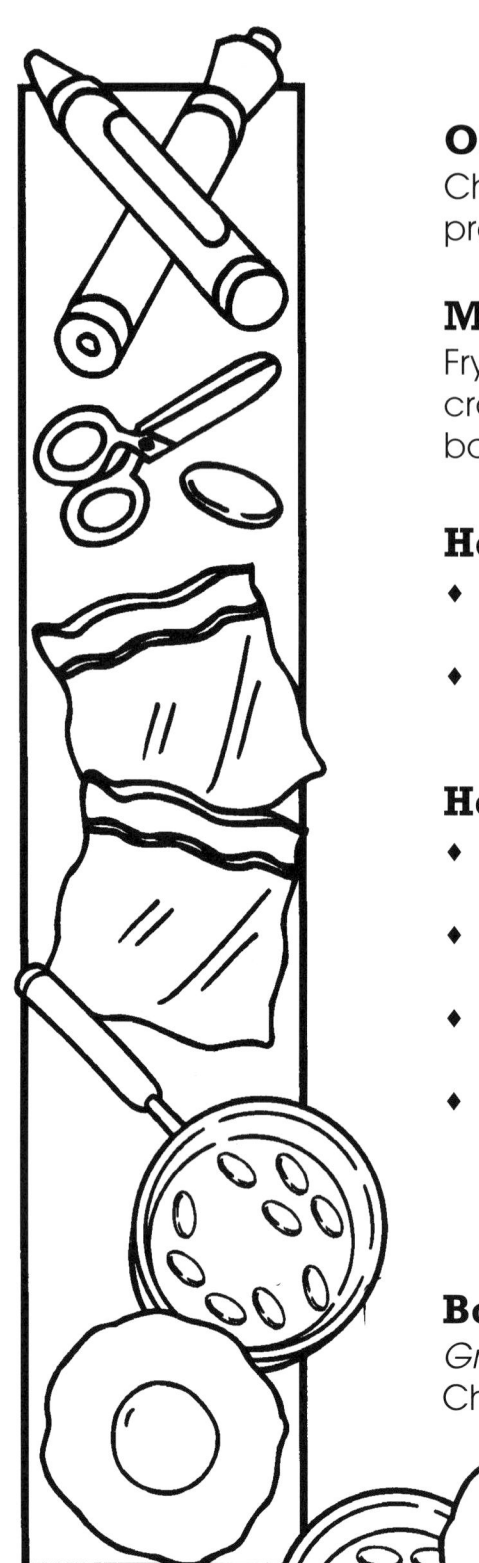

Objective:
Children will practise solving simple addition problems.

Materials:
Frying pans (page 123); Fried eggs (page 124); crayons or markers; scissors; counters; 2 resealable bags.

How to make the game:
- Duplicate the frying pans and fried eggs, colour, laminate and cut out.
- Store the frying pans and fried eggs in one resealable bag and the counters in another.

How to play the game:
- Take the frying pans out of the bag and line them up.
- Use the counters to help you solve the addition problems on each frying pan.
- Match each frying pan with the egg that has the correct answer on it.
- When you have finished, ask your teacher to check your work. Then place all frying pans and eggs back in one bag and the counters in the other.

Book link:
Green Eggs and Ham by Dr Seuss (HarperColin's Children's Books, 1980). Full of crazy drawings and silly rhymes

Frying pans

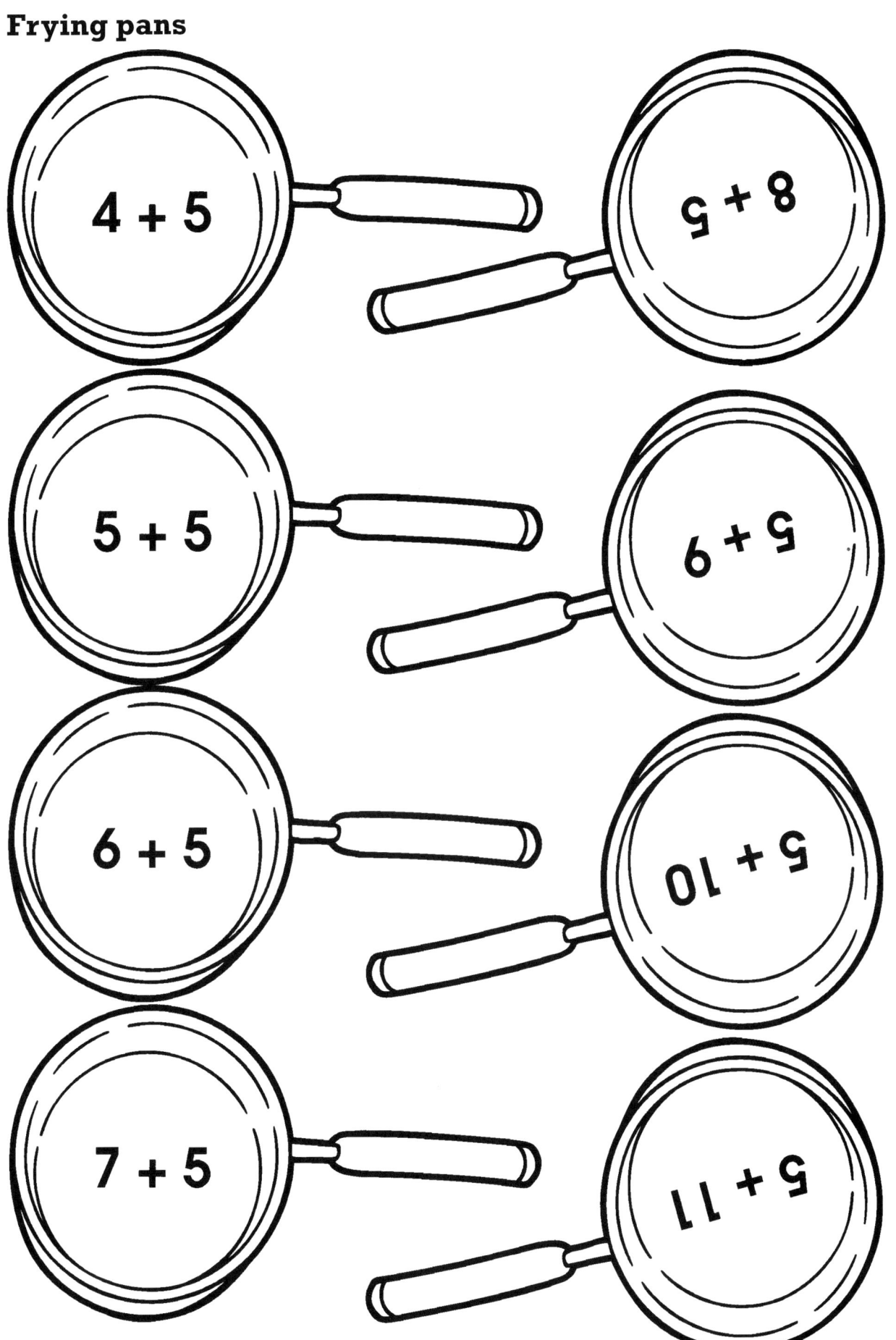

Fried eggs

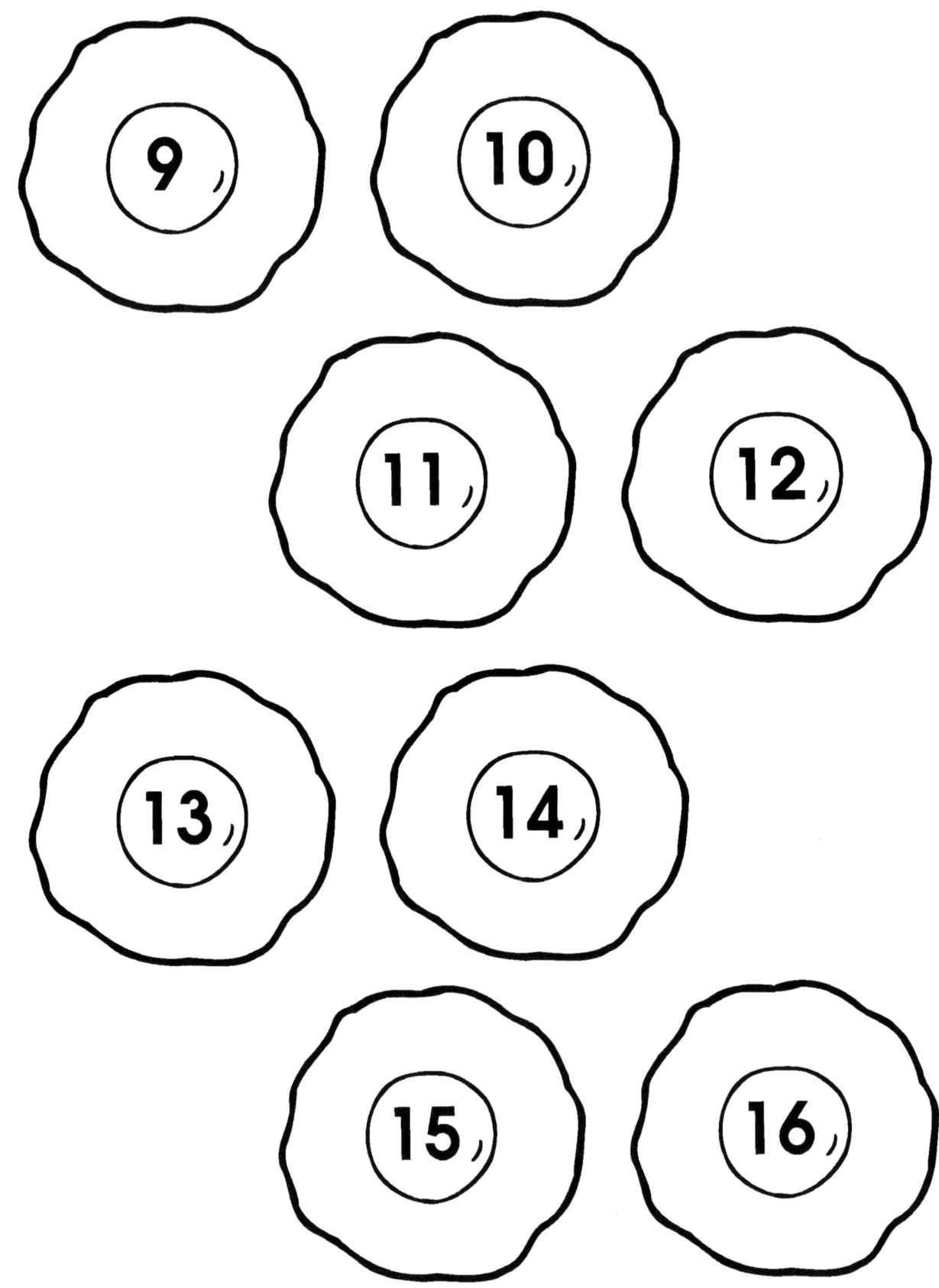

Brilliant Publications – www.brilliantpublications.co.uk
A – Z Maths Games by Karen M. Breitbart

Zoo animals

Objective:
Children will practise pairing matching numbers.

Materials:
Animals (pages 126–7); crayons or markers; scissors; 10 empty fruit punnets; index cards; tape; resealable bag.

How to make the game:
- Duplicate the animals, colour, laminate and cut out.
- Label each punnet with a number word from from one to ten using the index cards, markers and tape.
- Stack the punnets to store.
- Store the animals in a resealable bag and place the bag in the top punnet.

How to play the game:
- Take the animals out of the bag and look at the number written on each one.
- Line the animals up in a row from 1 to 10
- Line the punnet 'cages' in a row from one to ten.
- Place the correct animal in a 'cage' by matching the numbers on the animals to the numbers on the 'cages'.
- When you have finished, place the animals back in the bag. Stack the punnets. Place the bag of animals in the top punnet.

Book links:
If I Ran the Zoo by Dr Seuss (Random House, 1950). Packed with hilarious pictues and tongue-twisting rhymes.

Animals

Brilliant Publications – www.brilliantpublications.co.uk
A – Z Maths Games by Karen M. Breitbart

Animals

Skills and concepts covered

Brilliant Publications – www.brilliantpublications.co.uk
A – Z Maths Games by Karen M. Breitbart